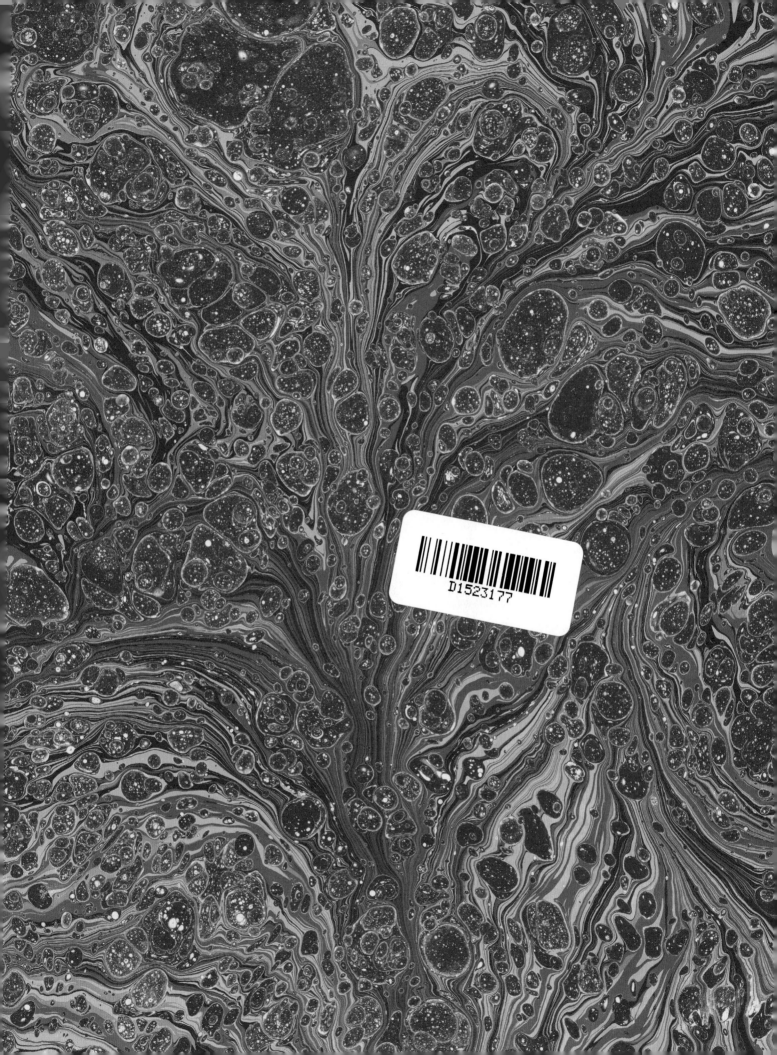

A Postcard History
of Japanese Aviation

昭和十三年十月二十七日

漢口陥落記念

A Postcard History
of Japanese Aviation

1910 – 1945

Edward M. Young

Schiffer Military History
Atglen, PA

Dedication

IN MEMORY OF YOSHIO AND CHIEKO TAGAYA

Book design by Stephanie Daugherty.

Copyright © 2012 by Schiffer Publishing.
Library of Congress Control Number: 2012932443

Printed in China.
ISBN: 978-0-7643-4039-0

We are interested in hearing from authors with book ideas on related topics.

Published by Schiffer Publishing Ltd.
4880 Lower Valley Road
Atglen, PA 19310
Phone: (610) 593-1777
FAX: (610) 593-2002
E-mail: Info@schifferbooks.com.
Visit our web site at: www.schifferbooks.com
Please write for a free catalog.
This book may be purchased from the publisher.
Try your bookstore first.

In Europe, Schiffer books are distributed by:
Bushwood Books
6 Marksbury Avenue
Kew Gardens
Surrey TW9 4JF, England
Phone: 44 (0) 20 8392-8585
FAX: 44 (0) 20 8392-9876
E-mail: Info@bushwoodbooks.co.uk.
Visit our website at: www.bushwoodbooks.co.uk

Contents

陸軍航空本部貸下

南支を制壓する陸軍新輕爆撃機

POST CARD

Acknowledgments

I WOULD LIKE TO THANK SEVERAL PEOPLE WHO helped in the preparation of this book. Daniel Hagedorn, Senior Curator at the Museum of Flight, first suggested to me that my collection of Japanese aviation postcards might make an interesting book. Ms. Sarah Serizawa, a student at the University of Washington, did the translations of many of the captions. Mr. Yoji Kan, a dealer in Asian postcards, sought out Japanese aviation postcards for me on his regular trips to Japan and encouraged my work on this project. My good friend Osamu Tagaya has long been a source of support for my interest in the history of Japanese aviation. Over the decades Sam has been exceptionally patient and generous in sharing his extensive knowledge of Japanese military aviation with me. This book is dedicated to the memory of his parents, Yoshio and Chieko Tagaya. Yoshio Tagaya grew up in Tokyo, studied ordnance at the University of Tokyo, and served as an ordnance officer with the Imperial Japanese Navy during the last year of the war. In his post-war career he was for many years a representative of the Japanese Government at several international commercial aviation organizations. As a boy, and later during his years at university, Mr. Tagaya collected aviation postcards. Over the years I had the pleasure of visiting with the Tagayas at their house in Zushi, southeast of Tokyo. Mr. Tagaya kindly shared with me his memories of his collecting activities and his knowledge of Japanese aviation postcards. He, too, encouraged my collecting activities.

The Postcard in Japan

FOLLOWING THE MEJI RESTORATION IN 1868, Japan embarked on a path of rapid modernization with the goal of preserving Japan's independence in the face of Western encroachment across Asia. Modernization affected nearly every area of Japanese society: Government administration, law, the military, the national education system, agriculture, transportation, and industry. The government sent representatives to Europe and the United States to study and recommend systems and practices that would enable Japan to become a modern state. A postal system was introduced in Japan in 1871, based on the British postal system. A few years later, in 1873, the Japanese Government issued the first postcard in Japan in order to encourage people to use the newly established postal system. Cheap and simple to use, postcards rapidly became popular, and by 1887 comprised the largest category of mail in the Japanese postal system.[1] Picture postcards had just begun to be issued in various countries in Europe. For the next 27 years the Ministry of Communication maintained a monopoly over the issuance of postcards, which it did not relinquish until October 1900. Private publishers then began issuing picture postcards. Within a few years postcards in Japan would enjoy a boom in popularity that would last for several decades. The postcard became a vehicle for various segments of Japanese society to convey a new and more modern image, a factor not lost on the Japanese military.[2] The postcard "...made possible the wide and rapid circulation of an

9

exciting, innovative visual vocabulary that had developed in response to broad debates about Japanese national identity and power..."[3]

For a population freed from the Tokugawa era restraints on travel, the postcard became a means of conveying to friends and family images of one's travels at home and abroad, as well as a simple and easily stored souvenir of the journey. But what really spurred the postcard boom was the 1904-1905 Russo-Japanese War. During the war hundreds of thousands of young Japanese men were mobilized as soldiers and sailors. The postcard became an important means of communication between the soldiers and their families. A postcard gave space for a quick message and had the added benefit of not having to be opened by the military censors. The Japanese Government found that the postcard had great benefit as a vehicle for propaganda to build national pride and support for the war. Postcards from the Russo-Japanese War portray heroic images of Japanese soldiers, sailors, and their commanders, as well as commemorating Japanese victories in the conflict. During the war the Government issued commemorative sets of postcards which became extremely popular; private publishers issued similar sets as well. The total number of postcards issued during the war may well have been close to one million.[4] The boom in the use of the postcard and in postcard collecting continued well into the 1930s, paralleling a similar postcard collecting fad in Europe and the United States. Postcards appeared covering a broad range of topics and scenes using photographs and western and traditional Japanese art forms.

In the first decades of the Twentieth Century postcards were "...symbols and vehicles of modernity."[5] For much of this era there was perhaps no better symbol of modernity than the airplane. Man's conquest of the age-old dream of flight stirred the imagination of millions around the world. Wilbur Wright's dramatic flying demonstrations in France during August 1908, his subsequent flights around the capitals

of Europe, and the great Grande Semaine d'Aviation de la Champagne at Reims, France, in August 1909 triggered widespread interest in this newest of man's arts. The aviation historian Charles Gibbs-Smith wrote "Reims marked the true acceptance of the airplane as a practical vehicle, and as such was a major milestone in the world's history."[6] Postcards in the hundreds captured images of this brave new world showing the pilots–Wright, Farman, Bleriot, Curtiss, Paulhan, Latham, and others–their airplanes and scenes of their triumphant flights. The airplane spread across Europe, the United States, and to Asia.

Japanese airplane postcards date from this early period. Airplane postcards began to be published in Japan soon after Imperial Japanese Army Captain Yoshitoshi Tokugawa made the first airplane flight in Japan on December 19, 1910. From then on postcards documented Japan's aviation progress. Given the ubiquity of postcards, it is not at all surprising that airplanes would quickly become a popular topic. As in other countries, Japanese aviation postcards illustrated the latest advancements in airplane design, celebrated flights, aviation's progress in other countries, and the practical application of the airplane in the military services and in transportation. From time to time the Government would issue special postcards to commemorate Imperial Japanese Army or Navy Day, or a special review of the fleet, and perhaps unique to Japan, to mark the occasion of presentation aircraft gifted to the military through public subscriptions. The pictures of Japanese pilots flying Japanese-made airplanes could not but help to bolster national pride and create an image of Japan as one of the world's leading nations, equal to the West. During the decade of the 1930s, as Japan became more militaristic and expansionist, postcards became again a vehicle for official propaganda and the celebration of the nation's military victories. For the young aviation enthusiast, airplane postcards were highly desirable and eagerly sought after to add to a collection or to paste in a scrapbook. Postcards were readily available from newsstands, stationery stores, and through the Post Office system for officially issued postcards. In the Imperial Japanese Army soldiers could purchase packs of 100 military-themed postcards for 25 Sen, equal to about 2% of a Superior Private's monthly pay.[7] Families in Japan could, within the bounds of military censorship, see the airplanes their soldier and sailor sons were flying. From the early days of flying in Japan up until the last years of World War II, when paper became scarce and strictly rationed, airplane postcards appeared regularly.

The Postcard as Propaganda

As many of the warring nations did in the years leading up to and during World War II, Japan made extensive use of postcards as a vehicle for propaganda aimed at both a domestic and a foreign audience. The postcard was an inexpensive, easily distributed means of conveying images and content supporting attitudes the Government wanted the Japanese people to adopt. Military propaganda was "...the primary motivation in the revival of picture postcards during the war years of 1937 to 1945."[8]

During the decade of the 1930s, Japanese society and Government came under the sway of an expansionist, ultranationalist military which saw Japan's very survival as under threat from the Western powers and the Soviet Union. From this viewpoint, the key to the nation's survival was the expansion of Japan's control over the Asian mainland and the removal of Western Imperialism from the rest of Asia to ensure that Japan had unfettered access to the raw materials the nation would need for its own defense. As a justification for following a path that would inevitably lead to war, the military needed an epic narrative to mobilize and motivate the Japanese people. The military and its supporters sought first, to inculcate in Japanese society a belief in Japan's spiritual and racial uniqueness,

or *kokutai* (national polity); second, to create within the people a deep-seated gratitude for being a part of the *kokutai* and a resolve to protect the *kokutai* through unquestioning obedience to its personification in the Emperor; and third, to persuade the people that the *kokutai* was under dire threat from the West in order to justify a program of military expenditures and territorial expansion. There was an equally strong sense that Japan had a sacred mission to liberate the rest of Asia from the yoke of Western control, and that Japan was, by right and by its own achievements, the modern leader of all Asia.[9] The function of the Japanese media, in all its forms, was "...to produce images and texts orienting citizens within the national cause as participants in a glorious history-in-progress one whose goals are lofty (in the case of Japan, freeing Asia from the grip of rapacious European colonial rulers and fulfilling the 'destiny' of the Yamato race and their ancient imperial tradition) and above all else, one that leads ineluctably to victory."[10] The goal of the propaganda campaign was to unify the citizens of Japan with the Japanese military in support of Japan's overseas expansion, bringing the home front and the battle front together as one.[11]

Picture postcards were "...among the most important sources presenting a wide variety of unofficial and official images..." during the 1930s and the early years of the Greater East Asia War, as the Japanese named it.[12] Airplanes presented an image of an industrial, technologically advanced nation, as well as showcasing Japan's military might for both a domestic and an international audience. For the Japanese citizen, these images were intended to instill a sense of pride in the nation's achievements. Once Japan was at war with China, and later with the Western powers after December 8, 1941, postcards showed images of Japanese Army and Navy airplanes in action to stir patriotic fervor. For an international audience in the rest of Asia, pictures of Japan's success in aviation helped convey an image that Japan was "...the most modern, the most advanced, and the strongest nation in Asia.”[13] As an example of the Government's use of postcards as propaganda, for Army Day on March 10, 1941, the War Ministry issued 350,000 postcards showing military subjects.[14] During the war postcards were a vehicle for spreading propaganda throughout the Greater East Asia Co-Prosperity Sphere. Postcards showing images of the Japanese military were translated into local Southeast Asian languages. Postcards also became a means of distributing examples of war art the Japanese Army and Navy commissioned during the war showing Japan's victories over the Allies.

The Cards

The postcards in this volume are all from my own collection, built up over nearly three decades of collecting. Building model airplanes in high school and my father's participation in the Pacific War as an intelligence officer with the Twentieth Air Force led me to develop an interest in the history of Japanese aviation. I first encountered Japanese aviation postcards in 1984 on a business trip to Tokyo. One weekend my good friend Osamu Tagaya took me to Kanda, Tokyo's bookstore district. Out of curiosity I asked him if there were postcards of Japanese airplanes available. A few inquiries brought us to a shop that sold a wide range of Japanese paper ephemera. The owner confirmed that they did indeed have aviation postcards, and pulled down a box for me to peruse. To my amazement there were postcards dating back to the early days of Japanese aviation, airplanes from the 1920s and 1930s, and even some from the early days of World War II. That afternoon's find was the beginning of my collection. Over the years I have continued the search in several cities in Japan, Europe, and in the United States. I was fortunate to have the opportunity to make repeated business trips to Japan over the next twenty years and to live in Japan from 1992 to 1994. The Kanda district became a regular haunt

whenever I had the time. In my halting Japanese I would ask book stores if they had aviation postcards. In most cases the answer was no, but occasionally I would find a store that had a small stock. Living in Japan gave me the opportunity to visit small antique and collectible fairs held at temples around the Tokyo area, and to attend some of the larger antique fairs in Tokyo and Kyoto. I managed to find postcards of Japanese airplanes at the Marché aux Puce in Paris, postcard shows in London, and through dealers in the United States. The advent of eBay opened up another source. In building this collection I have experienced the joys and frustrations that every collector knows. Months would go by when I would not find a single new card, and then a rarer card or a small collection would turn up unexpectedly. I remember most a July day in Tokyo during the rainy season. The rain was bucketing down, but for some reason I decided to go to the Kanda district during my lunch break despite the weather. At one of my regular stores I found 60 postcards from the World War II years which I had never seen before, or since.

For this volume I have selected postcards from my collection that are representative of the four main periods of Japanese aviation during the years 1910 to 1945. These cards comprise about a third of my total collection. Never having seen any listings of Japanese aviation postcards, I cannot say whether my collection is a modest or extensive one. I believe that in the period during which I have been collecting I have managed to assemble a collection that most likely mirrors the type and number of postcards produced during the 1910-1945 years, but I do not know for certain. There are certain puzzles about the postcards I have acquired that I hope I can one day address. Why, for example, are there far more postcards available of Imperial Japanese Army airplanes than Imperial Japanese Navy airplanes? Why are there comparatively few cards of the Zero fighter, which was such an iconic aircraft at the time and to this day? The relative paucity of postcards from the later war years can almost certainly be attributed to the declining availability of paper, but this, too, remains a question. It is my hope that the publication of this volume may lead other collections to surface and perhaps encourage more definitive studies of this media.

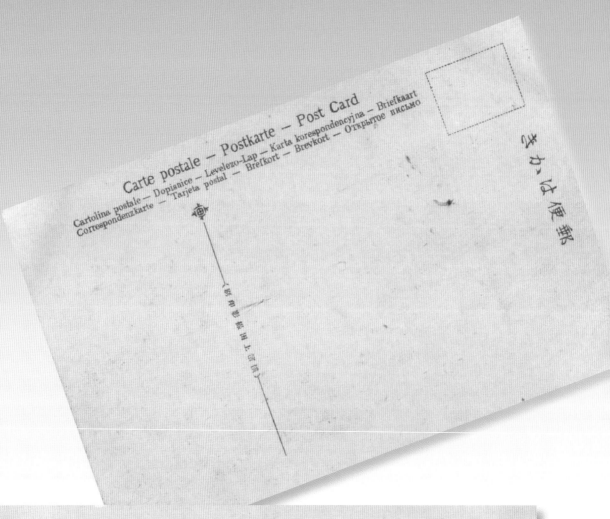

Carte postale — Postkarte — Post Card — Briefkaart
Correspondenzkarte — Dopisnice — Levelezo-Lap — Karta korespondencyjna
Cartolina postale — Tarjeta postal — Brefkort — Brevkort — Открытое письмо

(東洋飛行商會發行)   Narahara Biplane Otori.   奈良原式飛行機鳳號

CHAPTER ONE

The Pioneering Years

1910 – 1918

The Beginning of Aviation in Japan

REPORTS OF EARLY AIRPLANE FLIGHTS IN THE West had stirred the interest of several Imperial Japanese Army and Navy officers in the possible military applications of flight. An officer on the Imperial Navy's General Staff, Lt. Commander Eisuke Yamamoto–considered the conceptual father of Japanese Naval aviation–proposed in a paper that in the future naval warfare would incorporate the airplane.[1] The Navy Ministry entered into negotiations with the Army to initiate a program for the study of the potential of the airplane for military applications. On July 30, 1909, a few weeks before the Grande Semaine d'Aviation at Reims, General Sekiki Terauchi, Japan's Minister of War, established the Provisional Committee for the Study of the Military Application of Balloons (later the Provisional Military Balloon Research Association), comprised of Army and Navy officers and representatives from Tokyo Imperial University.[2] Despite its name, the Provisional Committee's main interest was in airplanes. Shortly thereafter, the Army sent Captain Yoshitoshi Tokugawa to France and Captain Kumazo Hino to Germany to learn to fly, and to purchase airplanes to bring back to Japan. The Provisional Committee then made plans to set up an airfield at Tokorozawa, in Saitama Prefecture, some 18 miles southwest of Tokyo.[3]

Tokugawa and Hino returned to Japan in November 1910, bringing with them a Henri

Farman biplane from France and a Grade monoplane from Germany. The two officers set up their airplanes in tents on Yoyogi Parade Ground in Tokyo and began preparations for a series of public flights. On December 19, 1910, Captain Tokugawa took off in his Farman at 7:55 in the morning for a flight of four minutes, circling the parade ground to thunderous applause from the huge crowds who had gathered to witness the first flight of an airplane in Japan. That afternoon Captain Hino made a flight of 1 minute and 20 seconds in his Grade. In April 1911 the Army's neophyte aviators moved to Tokorozawa, adding a Bleriot monoplane and a Wright Flyer built in Germany to their small stable of airplanes. Tokugawa and Hino flew these airplanes regularly during the rest of the year, often with large crowds in attendance.[4]

Two American pilots made notable flights during these early years of Japanese aviation. In early March 1911 Thomas Baldwin arrived in the Japanese city of Kobe with fellow pilots J.C. "Bud" Mars and Tod Shriver, dubbed the "Baldwin Flying Circus", to put on a series of exhibition flights in Japan following a successful display in Manila. Sponsored by the *Osaka Asahi* newspaper, the flights took place on March 11, 1911, at the Joto Parade Ground in Osaka before tens of thousands of spectators. At 1:20 in the afternoon on a beautiful day with light winds, J.C. Mars took off in Baldwin's biplane patterned after the Curtiss biplanes of the time. The *Japan Times* reported "...at a signal given, Mr. Mars got on his favorite biplane of the Baldwin model, and after a run of about 60 feet, the machine left the ground quite lightly and smoothly, and gradually rose to 200 feet. Then the aviator, keeping this height, described a circle about 2,200 feet in diameter; in a beautiful manner repeating this circular flight five times, to the great admiration of the spectators below."[5] Mars made three flights that day, and apparently went on to make flights at Kobe, Nagoya, and Tokyo in the days that followed. A little over a year later W.B. Atwater,

who had learned to fly at the Curtiss Flying School at North Island, made the first flight of a hydro-aeroplane in Japan. On May 10 and May 11, 1912, Atwater made several flights from the seashore at Yokohama, south of Tokyo on Tokyo Bay. Early on the morning of May 10 Atwater took off in his Curtiss hydro-plane and flew out over Yokohama harbor, circling the steamer *Tenyo* several times from a height of several hundred feet. The next day Atwater made three short flights in the afternoon, flying out over Yokohama harbor before thousands of spectators and several hundred Imperial Navy officers. His Curtiss hydro-plane was decorated with Japanese and American flags. On his second flight Atwater dropped a message from the Minister of the Navy to a senior Japanese admiral on board the torpedo-destroyer *Yayoi*. A few weeks later, on June 1st, Atwater made the first air mail flight in Japan, taking off from Shibaura on Tokyo Bay carrying 1,000 commemorative postcards to Yokohama, covering the distance in 29 minutes. On his return flight to Tokyo Atwater took back 1,000 postcards from Yokohama.[6]

Frustrated at the Army's lack of interest in hydro-airplanes, the Navy set up its own research body, the Committee for the Study of Naval Aeronautics, in 1910, beginning a bitter rivalry in aviation between the two services that would continue through World War II.[7] The Navy then sent its own officers overseas to learn to fly and to acquire aircraft; Lieutenants Kanehiko Umekita and Masahiko Kohama going to France to join Lt. Yozo Kaneko, and Lieutenants Sankichi Kono, Chuji Yamada, and Chikuhei Nakajima to the Curtiss Flying School at Hammondsport, New York. The Navy established its first naval air station at Yokosuka, near the western mouth of Tokyo Bay, in October 1912. That same month Lt. Kaneko returned from France, bringing with him a Farman hydro-plane. Lt. Kono was also recalled from the Curtiss school, returning with a Curtiss Model E hydro-aeroplane. These two pilots made their first flights in Japan on November 2, 1912, marking the start of Japanese naval aviation. Ten days later the Imperial Navy held a grand naval review off Yokohama with Emperor Taisho in attendance. The Navy's two new aircraft featured prominently in the review. Lt. Kono flew first, taking off in his Curtiss from the beach at Yokohama and making two large circuits of the assembled fleet before the Emperor in a flight lasting 14 minutes. Lt. Kaneko flew his Farman up from Yokosuka, arriving shortly after Kono, and also made two circuits over the fleet before his return flight, much to the delight of the spectators who lined the shore.[8]

The Development of Army Aviation

In addition to making regular flights, including the first cross-country flight in Japan, Captain Tokugawa designed the first military airplane built in Japan, an improved version of the Henri Farman biplane designated the Kaishiki No.1 Aeroplane (Kai-1) under the auspices of the Provisional Military Balloon Research Association, which designed and built a number of experimental aircraft types for the Army during this early period. The Kai-1 was built as a sesquiplane, with ailerons on the upper wings and a single horizontal tail surface.[9] The airplane made its first flight in October 1911. Following this success, during 1912 Captain Tokugawa supervised the design and construction of three more Farman-type airplanes, the Kai-2, -3, and -4 powered by 50hp Gnome or Anzani rotary engines. That same year the Army initiated training for pilots and observers. Two Army lieutenants sent to France in 1912 returned to Japan in the spring of 1913, bringing with them a 1912 Maurice Farman biplane, the M.F.7, and a Nieuport NG monoplane. Impressed with the Farman's performance, the Army ordered four more from France and decided to manufacture the type in Japan. The Farman type biplane became the mainstay of Army aviation for the next several years. The initial version was designated the Army Type Mo (Maurice Farman

Type) 1913; two similar airplanes were built as the Kai-5 and Kai-6. In 1915 the Army introduced an improved design of the Henri Farman 1914 model, the M.F.11, as the Army Type Mo-4 (Army Henri Farman Type Model 4). Later in 1916 the Army began building the Army Type Mo-6, which was similar to the Mo-4 but with a more powerful 100hp Daimler-type engine.[10] The Farmans became the mounts for the Army's first classes of pilots and observers. Farmans made a number of record flights around the country and engaged in Japanese Army aviation's first experience of combat.

At the outbreak of World War I, Japan joined the Allied powers and declared war on Germany based on its commitments under the Anglo-Japanese Alliance. Britain asked Japan to remove the threat of German commerce raiders in the Pacific and along the China coast. A combined Japanese and British force launched an attack against the German port of Tsingtao (now Qingdao), the base for the German Navy's East Asia Squadron. The siege of Tsingtao lasted from September 4 to November 16, 1914. In support of the Japanese Army units conducting the siege, the Army sent four Army Type Mo 1913 aircraft and its single Nieuport NG to a hastily built airfield near Tsingtao. From this field the Army planes flew their first mission on September 21, 1914. Two Farmans joined the Nieuport on reconnaissance over the German positions, dropping three small bombs from an altitude of 2,300 feet and returning with numerous bullet holes in the fabric of their wings.[11] Over the next six weeks the Army airplanes flew 86 reconnaissance and bombing sorties, dropping 44 15kg bombs from under wing racks on the Farmans. On October 13, 1914, the Japanese military's first aerial battle took place when several Army airplanes (and Navy airplanes–see the Navy section below) attempted to intercept the single serviceable German airplane at Tsingtao, a Rumpler Taube, without success. To match the German Taube, the Army hurriedly purchased two civilian Rumpler Taube airplanes that had been imported to Japan. One of these aircraft was rushed to the battlefront, but was damaged in the course of its first sortie just before the Germans surrendered.[12]

This brief experience of combat spurred the Army to expand its fledgling aviation arm. In December 1915 the Army organized the First Army Air Battalion, removing its flying unit from the Provisional Military Balloon Research Association. The First Army Air Battalion sent airplanes to the annual Army maneuvers in November 1916 and November 1917, where they were used primarily for reconnaissance by both sides in the maneuvers. In the fall of 1917 the Army established a second airfield at Kagamigahara in Gifu Prefecture, north of Nagoya, and formed the Second Army Air Battalion. The two Air Battalions were equipped with the Type Mo-4 and the Type Mo-6. The Army continued the development of experimental aircraft during the World War I period through the Provisional Military Balloon Research Association. Notable were the Seishiki-1 airplane, a tractor biplane built in 1916 reflecting developments in Europe and incorporating a more powerful engine, and the Seishiki-2, which flew in early 1918.[13] The Army also sent observers to Europe and the United States to report on the European air war and progress in aeronautical development. This led to the purchase of several aircraft from England, France, and the United States. During 1916, the Army purchased a Grahame White fighter and a Grahame White five-seater airplane, and in 1917 the Provisional Military Balloon Research Association purchased two Standard H-3 airplanes, building an additional three airplanes in Japan for use as training aircraft.[14] The Army's major purchases during these years, when it sought more modern aircraft, were from France, beginning in 1917 with the purchase of several Nieuport 24C fighters. In early 1918 the owner of one of the leading shipping companies in Japan donated half a million Yen to the Army and the same amount to the Navy to purchase more modern airplanes. Using this money the Army purchased 20 Sopwith 1 ½ Strutters built

in France by the Loiré et Olivier Company; 18 more Sopwiths were later built under license at Army arsenals.[15] In 1918 the Army purchased a small number of SPAD 7 aircraft and a single SPAD 11.[16] When in July 1918 Japan decided to send a large contingent of troops to join the Allied Expeditionary Force in Siberia in support of the White Russian armies against the Bolsheviks, the First and Second Army Air Battalions sent eight Type Mo-4 aircraft, six Type Mo-6 aircraft, and nine Sopwith 1 ½ Strutters to Siberia to support the Japanese Army's 12th Division.[17] At the end of World War I the Type Mo-4 and Type Mo-6 remained the predominant aircraft type equipping the First and Second Army Air Battalions alongside the small number of Sopwith, SPAD, and Nieuport aircraft.

The Development of Naval Aviation

Following its first successful flights in November 1912, the Imperial Japanese Navy decided to concentrate on the development of seaplanes, then the predominant type of naval aircraft among the European navies. In 1913 the Navy purchased two more Maurice Farman seaplanes and, like the Army, designated its Farmans the Type Mo Small Seaplane. The Navy began training more pilots with its four Farmans and Curtiss Model E Hydro-aeroplanes. The Navy's arsenal at Yokosuka began building both the Type Mo Farman and several Curtiss Model Es. That same year the Navy purchased its first naval aviation vessel, the *Wakamiya Maru*, a freighter converted to deploy and service the Farman seaplanes. This vessel, with several Farman seaplanes aboard, participated in the annual naval maneuvers that year. The first Japanese-designed airplane also appeared in 1913. Lt. Chikuhei Nakajima, who had studied at the Curtiss factory at Hammondsport, built an experimental seaplane patterned after the Curtiss Model E Hydro-aeroplane that made several successful flights, though it was not adopted. In 1914 the Navy ordered a Deperdussin float monoplane and a Maurice Farman 1914 seaplane, a larger version of the 1912 model with a more powerful 100 hp engine that could seat a pilot and two observers. This became the Type Mo Large Seaplane. The Navy's new seaplane arrived just before the Navy's small air unit was called on to participate in the attack on Tsingtao.[18]

In late August 1914 the *Wakamiya Maru* was assigned to the Navy's Second Fleet and sent out to join the blockade of the German naval base at Tsingtao with three Type Mo Small Seaplanes and the newly-arrived Type Mo Large Seaplane. On September 5th, Lieutenant Hideho Wada in the larger three-seat Farman seaplane and Sub-Lieutenant Masaru Fujise in a smaller Type Mo took off on the first Japanese air mission. Flying over the German fortifications, the two pilots dropped several small bombs made from naval shells which missed their target, but performed a valuable reconnaissance over the nearby Kiao-Chou Bay, identifying the German ships still at anchor. When the *Wakamiya Maru* was damaged by a German mine the four seaplanes moved to a stretch of beach, from where they continued to fly reconnaissance and bombing missions, including missions scouting for floating German mines. On September 24th, Lt. Wada and Lt. Yamada, flying Type Mo Small Seaplanes, attacked a German destroyer, the Japanese Navy's first aerial attack on an enemy vessel. On October 13th, Lt. Wada took off in his Farman to join the three Army aircraft attempting to intercept the German Taube, which escaped. During the brief campaign the Navy seaplanes flew 49 sorties, dropping nearly 200 bombs. While the results of these flights were by no means dramatic, the airplane had proved to be invaluable in reconnaissance, a fact that was not lost on the Navy's high command.[19]

Although the Navy's airplanes saw no more combat service during World War I, the Navy, like the Army, continued to develop its aviation arm. The Navy trained more pilots, developed experimental aircraft and purchased some from abroad, and organized a more permanent

structure for naval aviation. In early 1916 the Navy received funding to establish permanent air groups, designated *Kokutai*.[20] The first air group was organized on April 1, 1916, at Yokosuka, taking its name from its station as the Yokosuka Air Group (*Yokosuka Kokutai*). The focus of this first air group was on training new pilots and testing new naval aircraft. The Yokosuka Naval Arsenal had begun production of the Type Mo Large Seaplane in 1915, building some fifteen aircraft re-designated the Ro-go Otsu-gata.[21] Between 1916 and 1918, Lt. Chikuhei Nakajima built three experimental seaplanes, including an experimental twin-engine seaplane capable of carrying a torpedo. In 1917 Nakajima's assistant, Lieutenant Kishichi Umakoshi, designed a reconnaissance seaplane that had superior performance to any of the Navy's current airplanes. First flown in early 1918, the Navy ordered the type into production as the Yokosho (for Yokosuka Naval Arsenal) Ro-go Ko-gata Reconnaissance Seaplane.[22] Powered by a 200 hp Hispano-Suiza E engine built by Mitsubishi, the Ro-go Ko-gata replaced the older Farman seaplanes and remained in service until 1926, the first Japanese naval aircraft built in quantity. As the Army had turned to France and the United States for newer aircraft, the Navy looked to England for naval aircraft. In 1916 the Navy purchased a single Short Brothers S.184 reconnaissance seaplane and a single Sopwith Schneider Fighter Seaplane which became the Ha-go Small Seaplane, the first in a type of aircraft the Japanese Navy would perfect in the coming years. The aircraft factory at the Yokosuka Arsenal built three S.184s, but the type was not put into production.[23] In 1918 the Navy purchased a single Tellier T.3 flying boat from France. This was the Navy's first direct experience with a flying boat, but the tests proved less than successful.[24] The Navy, too, sent observers to the United States and Europe to report on progress in naval aircraft development and the use of naval aircraft in the war. Having concentrated on seaplanes for several years, the British experiments with flying aircraft off the decks of naval vessels was a revelation; several Japanese Navy observers reported on the construction of the British aircraft carriers *Argus* and *Hermes*.[25] It was becoming evident that while naval aviation was proceeding apace Japan was falling behind.

The Development of Civil Aviation

The romance and adventure of flight appealed to many young Japanese men outside the military forces. In the pioneering era civil pilots were active in promoting aviation and building their own aircraft. Notable among these early flyers was Baron Sanji Narahara, who built the first successful airplane made in Japan, which carried out a short flight in May 1911. The Narahara No.4 airplane was completed in March 1912. Named the *Ohtori-go*, this airplane made one of the first paid exhibition flights in Japan, barnstormed around the country, and made flights in Korea and Manchuria. Between 1910 and 1919 more than 20 individuals or associations built and flew their own aircraft, though none were built in quantity. The Imperial Aeronautic Association was formed in 1913 through the amalgamation of two earlier associations to promote civil aviation in Japan. The IAA purchased two Rumpler Taube airplanes which were delivered in early 1914. Prior to being sold to the Army for use in the siege of Tsingtao, the Taube airplanes were used to make several exhibition flights and to begin instruction of Japanese pilots. The IAA also sponsored the First Civil Flight Competition on June 13-14, 1914. Five pilots competed in the event, which attracted 270,000 spectators.[26]

With no flying schools available in Japan, young men who wanted to learn to fly had to go overseas. A number went to France, but more went to the United States to train at the Curtiss flying schools at Hammondsport and at North Island, and at other schools across the country. Motohisa Kondo was the first Japanese to win a pilot's license in the United States, but sadly

died in the crash of a Curtiss pusher in Los Angeles on October 6, 1912, the first Japanese to die in an air crash. The second Japanese pilot to win a pilot's license in the U.S., Koha Takeishi, returned to Japan with a Curtiss pusher presented to him by Japanese residents. On May 4, 1913, the *Asahi Shimbun* newspaper sponsored Takeishi's flight between Osaka and Kyoto. Landing at a parade ground in Kyoto, Takeishi crashed and was killed in the first civil air accident in Japan. Still, despite these and many other incidents in the U.S. and Europe, the number of young Japanese learning to fly grew slowly but steadily. Juichi Sakamoto and Takayuki Takaso, who won prizes at the First Civil Flight Competition in 1914, both won their pilot's licenses in the United States. The 1915 Yearbook of the Aero Club of America listed 13 active Japanese members. The first flying school in Japan was established in 1917, when Seitaro Tamai and Tamotsu Aiba set up the Japan Flying School at Haneda, near Tokyo, using a biplane Tamai had designed.[27]

Before the United States became involved in World War I several American pilots came to Japan to give flying exhibitions, following in the footsteps of J.C. Mars and W.B. Atwater. At the end of 1915 Charles F. Niles brought a Bleriot monoplane and a Curtiss pusher to Japan and gave a series of demonstration flights over Tokyo, causing a sensation with his aerobatic maneuvers. In March 1916 Art Smith, the noted American stunt pilot, arrived in Japan for a series of exhibition flights. Flying a Curtiss pusher, Smith held several aerobatic shows in Tokyo during early April which surpassed Niles in daring stunts. Smith attached smoke devices to his airplane so that the crowds below could follow his maneuvers as he performed loops, spins, and a falling leaf. His shows were tremendously popular. Smith went on to give demonstrations around Japan until he broke his leg landing his plane at Sapporo, in Hokkaido. Smith returned to Japan the following year with equal success. At the end of 1916 Katherine

Stinson, like Art Smith a noted stunt pilot, came to Japan as part of an Asian tour. Stinson also proved quite popular, giving a number of air shows in Tokyo and Osaka.[28] Of her performance at Tokyo a local journalist reported, "Japan marveled at the feats of the late Charles Niles; gasped at the daring and mastery of Art Smith, but displayed nothing short of ecstasy yesterday afternoon when a nineteen-year-old member of the gentler sex, with her 100 odd pounds of pluck, demonstrated that she could duplicate the feats which have caused the death of one man, the serious injury of another and have been mastered only by a few persons in the world… When the second exhibition had been concluded, thousands surged toward the hanger, crying 'Banzai' as they ran, anxious for a glimpse of the first girl they had ever seen in the role of aviator."[29]

The pioneering era saw the tentative beginnings of the Japanese aviation industry. Several individuals attempted to set up companies to manufacture airplanes, notably Seitaro Tamai, who built a number of airplanes of his own design, but the one success from these early years was the firm that Navy Lieutenant Chikuhei Nakajima founded in December 1917 as the Hikōki Kenkyūshō (Aeroplane Research Institute). A trained engineer, Nakajima had studied airplane construction at the Curtiss factory at Hammondsport and had learned to fly there. Returning to Japan, he became the Navy's first aeronautical engineer. Nakajima believed strongly that the airplane should be the Navy's weapon of the future. In 1915 he sent a memo to the Navy arguing that the money spent on developing capital ships could be better spent developing airplanes, and that this could best be done through private companies and not military arsenals.[30] Nakajima resigned from the Navy in the summer of 1917 to found his own airplane company. In April 1918, in order to obtain more capital, Nakajima entered into a partnership with Seibei Kawanishi, the owner of the Nippon Wool Company in Kobe. Together they formed the Goushikaisha Nihon Hikōki Seisakusho (Japan Aeroplane Works Company Ltd.). That year Nakajima designed and built two airplanes, the Nakajima Type 1 and Type 3, using Hall-Scott engines that he obtained through his contacts with a senior Army officer and member of the Provisional Military Balloon Research Association. Though neither type went into production, this was the beginning of what would become in the decades ahead one of Japan's largest and most important manufacturers of military and commercial airplanes.[31]

Postcards from the Pioneering Years

式トイラ機行飛用軍本日大

Captain Kumazo Hino flying the Wright Flyer imported from Germany in the spring of 1911.

J.C. Mars and Tod Shriver in Japan, March 1911.

J.C. "Bud" Mars about to take off on a flight in March 1911. Tod Shriver stands behind holding the tail of Mars' Baldwin model pusher.

（乗坐氏スーム）號雲飛機行飛式葉復イウドルーボ國米

（寫實）景光行飛大上海式スチーカ　氏ーターォウツァア士名　ルケ於ニ濱大宮の西津撮

三其（景光ノ走滑上水）乗座氏ーターォウツァア機行飛式上水スチーカ

W.B. Atwater's Curtiss Hydro-aeroplane on the beach at Yokohama, May or June 1912.

W.B. Atwater taxis in to the beach at Yokohama after one of his flights.

尉大田山はるせ乗座（式スチーカ）の前庫納格道濱飛行軍海氣式橫
（行發社聞新央中�相）　員係機全は他其尉中瀨中央は尉中瀨

Lt. Chuji Yamada sits at the wheel of the Imperial Japanese Navy's Curtiss Model E Hydro-aeroplane at the first Naval air station near Yokosuka. Yamada flew this airplane during the grand naval review on November 12, 1912.

Army Aviation

號三第式川德及式ンマルアフスリーモ　　（機行飛用軍本日大）

The Kaishiki No.1 Aeroplane, left, and the Kaishiki No.2 Aeroplane at Tokorozawa in the summer of 1912. Patterned after the Henri Farman, the Kaishiki No.1 was the first military airplane built in Japan.

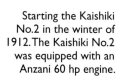

Starting the Kaishiki No.2 in the winter of 1912. The Kaishiki No.2 was equipped with an Anzani 60 hp engine.

景光ノ那刹ルストン七行飛ヤ今機飛式川德　　（機行飛用軍本日大）

Imperial Japanese Army 1ˢᵗ Lieutenant Oka sitting in the Kaishiki No.3 airplane at Tokorozawa in 1913.

徳川式第三號ニ乗シ正ニ出發セントスル岡中尉

（大日本軍用飛行機）　　七式飛行ニ乗リ岡中尉關東平橫斷大飛行出發ノ光景

Lt. Oka in a Maurice Farman 1913 (M.F.7), which became the Army Type Mo 1913 Aeroplane. The Type Mo was the Army's standard airplane during the pioneering years.

ニユーポルト式飛行機　（陸軍々用）

The Nieuport NG airplane. The Army purchased a single example in 1913. This airplane later flew at the siege of Tsingtao in 1914.

The Kaishiki No.7 Aeroplane. Lt. Shigeru Sawada modified a Henri Farman 1914 model, the Maurice Farman M.F.7, imported to Japan at the end of 1914, adding a slight sweep-back to the wings and replacing the Gnome engine with a Curtiss OX-5. The airplane first flew in April 1915.

陸軍々用　飛行機新造アンリーファルマ式飛行機

Parade ground of the Toyohashi. 豊橋練兵場

A color tinted postcard of the Army Henri Farman Type Model 4 Aeroplane (the Type Mo-4).

An Army Type Mo-4 coming in for a landing. Note the two other Type Mo-4 airplanes in the foreground and the canvas hangars in the rear.

The Seishiki-1 Aeroplane. In 1916 the Provisional Military Balloon Research Association began work designing a tractor biplane with a 100 hp Mercedes-Daimler engine the Imperial Aeronautic Association had imported from Germany to install in its Rumpler Taube. The Seishiki-1 made its first flight on May 6, 1916.

（從軍撮影）　某軍航空隊の活動　（陸軍特別大演習紀念）

New Hikoki the army

陸軍々用　戦闘用百馬力式飛行機

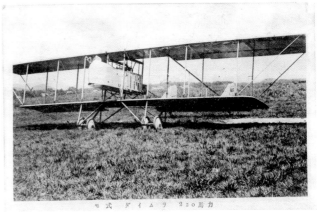

220馬力　ラムダイ式モ

The Army Maurice Farman Type Model 6 (Army Type Mo-6 Aeroplane), a more powerful version of the Type Mo-4 Aeroplane with a license-built 100 hp Daimler engine, based on the Maurice Farman M.F. 11. The Type Mo-6 was introduced in 1917 for reconnaissance and training. The Army purchased 134 between 1917 and 1921.

丁式一型襲撃機

The Standard H-3 Trainer. The Provisional Military Balloon Research Association purchased two Standard H-3s from the U.S. Army in May 1917. Despite being considered dangerous to fly, the imported Standards and three more built in Japan were used for training.

陸軍最新式ニユボールト式三ソツピー式駆逐用

The caption reads "The military's new type airplanes-Nieuport and Sopwith". In 1917 the Army purchased examples of the Nieuport 24C fighter and the Sopwith 1 ½ Strutter, license-built in France.

陸軍飛行機連射砲　（生活軍隊）

A postcard from the early 1920s showing the gunner's position on one of the French-built Sopwith 1 ½ Strutters.

戦闘用ソッピース式ローン百弐拾馬力

A tinted postcard also from the early 1920s showing a Sopwith 1 ½ Strutter, possibly one of the airplanes built in Japan.

Naval Aviation

One of the Navy's Curtiss Model E Hydro-aeroplanes taxis in after a flight over Yokohama Bay.

櫻行飛式ルるあつゝし走滑を上波　黄昇第七揖

Rear view of a Type Mo Small Seaplane on the landing ramp, ready for the next flight.

式ンマルアフ機行飛軍海るへ向ニ臺走滑

式ンマルアフ機行飛軍海ナアニ臺走滑

A Navy Type Mo Small Seaplane taxis up to the landing ramp. Three Type Mo Small Seaplanes flew at the siege of Tsingtao.

The Navy Type Mo Large Seaplane, based on the Maurice Farman M.F.11, had a larger 100 hp engine and could carry a crew of three.

（力馬百乗人三）式ンマルアフ機行飛軍海ルへ向ニ臺走滑

In 1914 the Navy imported a single Deperdussin twin-float monoplane.

A Type Mo Small Seaplane being hoisted onto the seaplane carrier *Wakamiya*.

A Type Mo Small Seaplane flies over the Japanese fleet lying off the German-controlled port of Tsingtao in the fall of 1914.

機行飛上水力馬十二百式スーピッソ
濟可許号四五七第鎭横

The Navy Ha-go Small Seaplane. The Navy purchased one Sopwith Schneider fighter seaplane in 1916. This was the Navy's first fighter plane.

力馬十四百二式廠横機行飛上水

The Navy's Yokosho Ro-go Ko-gata Reconnaissance Seaplane, built at the Yokosuka Naval Arsenal beginning in 1918. The Navy ultimately received 218 Ro-go Ko-Gata seaplanes, the Navy's first mass produced airplane.

帝 國 海 軍　練習用横廠式飛行機百十七号着水刹那ノ光景

A Navy Ro-go Ko-gata Reconnaissance Seaplane in flight.

A tinted postcard showing a Ro-go Ko-gata Seaplane taxiing out for a flight.

THE NAVY AEROPLANE

THE NA'VY A'EROPLANE 機行飛上水式廠工軍賀須横の中行飛

A tinted postcard showing a Ro-go Ko-gata Seaplane in flight over an Imperial Japanese Navy vessel. A Type Mo Seaplane flies in the background.

Civil Aviation

Tsuzuki, Monoplan. 機行飛葉單號二第式筑都（製霞許不）

Tesusaburo Tsuzuku was one of the first Japanese private citizens to design and build his own airplane. The Tsuzuku No.2 Airplane, first flown in the summer of 1912, made a number of demonstration flights for the military and for the public.

況實之行飛大諸鴨市都氏石武催主聞新日朝版大

Koha Takeishi was the second Japanese pilot to earn his Aviator's Certificate from the Aero Club of America, winning Certificate No.122 in May 1912 at the Curtiss Flying School in San Diego, California. On May 3, 1913, Takeishi took off in his Curtiss Model E Pusher, named *Shirohato* (White Dove), for a flight from Osaka to Kyoto. Landing at Kyoto his plane crashed and he was killed.

景光ノストセニ滑走大行飛酒樽乗氏武行新聞社主催阪每日日三月五

Koha Takeishi just before taking off from Osaka on his fatal flight to Kyoto.

Takayuki Takaso learned to fly in the United States, receiving Aviator's Certificate No. 219 from the Aero Club of America. He competed with this airplane in the First Civil Flight Competition in 1914, taking second place in the flying time contest.

行發館寫光愛市阪大　氏右左高縱操者機行飛號四第式右左高　（製復許不）

行發館寫光愛市阪大　氏右左高家行飛號九十二百二國米　（製復許不）

Takayuki Takaso in his airplane. The engine is a 60 hp Hall-Scott. Takaso made a number of demonstration flights in this airplane in cities around Japan.

野銀島藏氏座乘ノカーチス式飛行機

Genzo Nojima received his Aviator's Certificate No.199 from the Aero Club of America. In 1914, at the invitation of the Japanese Government he took his Glenn Martin Pusher on a barnstorming tour of Taiwan, then a Japanese colony.

In 1913, Lt. Commander Onokichi Isobe travelled to Germany to learn to fly the Rumpler Taube and to purchase two Taubes for the Imperial Aeronautic Association, returning to Japan in early 1914. Isobe is seen standing, left, with one of the IAA's first student pilots, Yukiteru Ozaki.

（大日本軍用飛行機）　ルムプラー正面左部礒少佐右尾崎氏

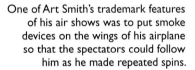

Art Smith was a noted American stunt pilot who spent several months in Japan in the spring of 1916 giving exhibition flights with great success.

One of Art Smith's trademark features of his air shows was to put smoke devices on the wings of his airplane so that the spectators could follow him as he made repeated spins.

The American aviator Charles Franklin Niles in the Bleriot he brought to Japan in December 1915. Niles' aerobatic flights over Tokyo were an outstanding success.

（米國航空界の女王スチンソン嬢）　日本服の艶姿

（空界の女王スチンソン飛行機）　ライト式復葉機の正面

Katherine Stinson was immensely popular in Japan. She stands next to her specially built Partridge-Keller biplane in a Japanese Kimono given to her by Lt. General Gaishi Nagaoka, President of the Imperial Aeronautic Association.

A tinted postcard showing the Partridge-Keller biplane with an 80 hp Gnome radial engine that Katherine Stinson used during her exhibition flights in Japan

（スミス飛行機）　カーチス九十馬力発動機装置復葉機の正面

Art Smith returned to Japan in 1917 for a second series of exhibition flights. This postcard shows the airplane he brought with him.

偵察機の飛行

POST CARD

The Apprenticeship Years

1919 – 1930

The Twenties:
The Decade of Internationalism

JAPAN EMERGED FROM WORLD WAR I AS ONE OF the Great Powers with a much stronger, more industrialized economy and a more prominent role in Asian affairs. The decade that followed was one of continued change in political affairs, the economy, in Japan's foreign relations, and within the military. At the end of 1926 the Emperor Taisho died. His son, Crown Prince Hirohito, became the new emperor, taking the reign name Showa. Japan's experience during the 1920s has led some historians to label this period of the nation's history the "liberal twenties".[1] A parliamentary democracy with active political parties predominated during this period, though with frequent changes of cabinet and prime ministers. In the conduct of its foreign affairs Japan was internationalist, having a permanent seat on the Council of the League of Nations and participating in deliberations with the other Great Powers on the critical issues of the day, particularly the limitation of armaments. Following a recession in the early 1920s Japan's economy resumed its growth, especially in iron and steel production and heavy industry, which nearly doubled by the end of the decade. The decade witnessed the expansion of both small-scale factories and workshops and the rise of the *zaibatsu*, the giant conglomerates with a broad range of activities in multiple industries.[2]

In the early years of the decade the Japanese Army and Navy were subject to cuts in their size and their budgets, though throughout the decade the two services managed to maintain a sizable share of government expenditures. The Navy's budget reached its lowest point in the 1925-26 budget cycle, the Army a year later.[3] Despite the cutbacks the Navy, in particular, was able to argue successfully that the newer technologies that came out of World War I, especially aviation and submarines, needed to be added to the Navy's capabilities.[4] There were cutbacks in strength as well. Under the Washington Naval Treaty Japan had to scrap ten naval vessels and accepted the 5:5:3 ratio of capital ships for the following ten years. In 1925 the Army lost four of its divisions to free up funds for modernization, which included an expansion of Army aviation.[5] Only toward the end of the decade did military budgets resume their growth.

For Japanese aviation the Twenties were the years of apprenticeship. After the end of World War I the Army and the Navy realized that the technological and tactical changes that had come out of the air war over Western Europe had left their small air arms woefully behind those of the other Great Powers. The Japanese Army and Navy had to learn how to apply the airplane for military use, not simply how to fly. Lacking its own established indigenous aviation industry and a cadre of trained aeronautical engineers and military airmen, to catch up to the West Japan needed to undertake an intensive period of instruction from the leading Western air forces and borrow heavily from Western aeronautical technology, which was well in advance of anything in Japan at that time.[6] The Japanese military decided that military aircraft and engines should be designed and built in the private sector and not in military arsenals.[7] During the 1920s, with the active encouragement and sponsorship of the military, Japan built up an indigenous aviation industry. The aircraft companies that emerged during this period entered into licensing agreements with foreign firms, hired foreign engineers to work in Japan, and sent large numbers of young Japanese engineers to study aeronautical engineering at European and American universities and to gain experience working with foreign firms.[8] As the decade progressed, the capabilities of the Japanese aviation industry expanded rapidly under this intensive program of instruction. By the end of the twenties Japanese airplane manufacturers were on the verge of independence from the West, as the Army and Navy had intended.

The Development of Army Aviation

The Army was the first of the two services to seek Western instruction. In 1918, having decided to purchase a large number of modern French military airplanes, the Army initiated talks with the French Army about the possibility of sending instructors to Japan to train Japanese Army pilots in their use. The result was the French Aeronautical Mission, which sent 61 instructors from all branches of the French Army under the command of Colonel Jacques Fauré. The French mission arrived in Japan in January 1919 and spent the next nine months training Japanese Army pilots and a select few Navy pilots in the latest techniques of aerial combat, gunnery, bombing, and reconnaissance at a number of Army air fields around Japan.[9] The French instructors began with a refresher course for all pilots and worked up to the latest tactics. The French mission was eminently successful in bringing advanced flying techniques to the Japanese Army's air units. French aircraft remained predominant in the Japanese Army well into the 1920s. The French mission brought with it 40 SPAD 13 fighters for instructing the Japanese in aerobatics, one or two Breguet 14 bombers, and several Salmson 2A2 reconnaissance airplanes. These aircraft joined other SPAD 7 and 13 and Nieuport 24 and 27 fighters, as well as Nieuport 81 and 83 trainers the Japanese Army had previously ordered from France, together with 50 Sopwith Pups ordered in 1919 from England. Impressed with the Salmson 2A2, the Army arranged for the newly-formed

Aeroplane Section of the Kawasaki Dockyard Company to build the type under license as the Type Otsu 1 Reconnaissance Aircraft. In 1919 the Army arsenal at Tokorozawa began building the Nieuport 24 as the Army's standard fighter (the Army's pilots preferring the Nieuport's superior maneuverability over the SPAD 13), later transferring production to the Nakajima Aeroplane Company as the Army Type Ko 3 Fighter/Trainer. In 1921 the Army decided to build the Nieuport 81 and 83 trainers in Japan, assigning production of the Nieuport 81 to the Mitsubishi Aircraft Company as the Army Type Ko1 and the Nieuport 83 to Nakajima as the Army Type Ko 2. Mitsubishi later took on license production of the French Hanriot HD-14 as the Army Type Ki-1 trainer. Completing the group of French aircraft adopted in the early 1920s, Nakajima was given a contract to build the excellent Nieuport 29-C-1 under license, building 608 of these airplanes for the Army from 1923 to 1932.[10]

In early May 1925, the Army's air corps was reorganized and made an independent branch of the Army on par with the artillery, cavalry, and infantry. At the same time the Army Air Battalions (*Hiko Daitai*) were also reorganized into larger Air Regiments (*Hiko Rentai*), which consisted of two Air Battalions comprised of two or more air squadrons (*Hiko Chutai*).[11] The Army's Air Regiments were a mix of homogeneous aircraft types, such as fighters or reconnaissance aircraft, or a combination of types. In 1926 the Army had eight Air Regiments consisting of two fighter regiments, four reconnaissance regiments, and two combined regiments with some 500 airplanes.[12] The Army's emphasis was on tactical operations, so within its eight Air Regiments the Army had only one squadron of heavy bombers equipped with the Army Type 87 Heavy Bomber, a joint development of the Kawasaki Aircraft Company and the Dornier Company with the assistance of seconded Dornier engineers (in 1927 the Army introduced a new classification system for its aircraft which

combined the airplane's function and the last two digits of the Japanese year in which the plane entered service, in this case 2587).[13]

As part of the 1925 reorganization the Army drew up expansion plans to increase the number of squadrons and the number of aircraft. In addition, the Army decided to replace its existing aircraft with newer models. The Japanese Army's first generation of post-World War I aircraft were license-built through tie-ups between the new Japanese aircraft manufacturers and established European firms, a necessary step in the education of the industry, but not one that would lead to ultimate autonomy. The next stage of development was to begin to design aircraft in Japan, and to this end in 1925 the Army initiated its first design competition in an effort to both stimulate and support the development of the Japanese aviation industry. The Army submitted a request to Nakajima, Kawasaki, and Mitsubishi to design a light bomber. The following year the Army requested the Ishikawajima Aeroplane Manufacturing Company, Kawasaki, and Mitsubishi to submit designs for a new reconnaissance aircraft to replace the Type Otsu-1; Nakajima also came up with its own entry with the help of two French engineers. All three companies chose to recruit German aeronautical engineers to assist them. The Army decided not to adopt any of the light bomber designs that came out of the 1925 competition, but Kawasaki's entry in the reconnaissance competition proved to be a winner. Adopted as the Army Type 88 Reconnaissance Aircraft, Kawasaki built 710 in two versions and an additional 407 were built as the Army Type 88 Light Bomber; both types served well into the 1930s. In 1927 the Army began its third competition, submitting requests to Kawasaki, Mitsubishi, and Nakajima, who were rapidly becoming its preferred manufacturers, to design a new fighter aircraft to replace the Nakajima Ko-4. The three prototypes that emerged from these firms were all parasol monoplanes designed with the assistance of German and French engineers. Though none were immediately accepted, Nakajima continued development of its entry, which was adopted a few years later as the Army Type 91 Fighter. By the end of the decade the Army had completed the expansion program for its air arm. The Army's eight Air Regiments had 26 component squadrons, of which 11 were fighter squadrons, 11 were reconnaissance squadrons, and four were bomber squadrons (two light bomber and two heavy bomber squadrons) totaling some 800 aircraft. While the French designed Nieuport 29-C-1 (Type Ko-4 Fighter) and Salmson 2A2 (Type Otsu-1 Reconnaissance) were still in service, an increasing number of Army aircraft were from the second-generation, designed in Japan with foreign assistance.[14]

The Development of Naval Aviation

Like the Japanese Army, the Imperial Japanese Navy realized after the end of World War I that its own small naval air arm was both woefully inadequate for the force the Navy wanted to build and behind the Western powers in both capability and doctrine. Like most other countries with a naval air force, the Japanese had concentrated on seaplanes for reconnaissance for the fleet, but Japanese Navy observers in England during the World War had reported on the British use of warships to fly off aircraft and the Royal Navy's work on the *Hermes*, Britain's first aircraft carrier.[15] This was an entirely new area for the Japanese Navy which the Navy was soon to exploit, laying down the Navy's first aircraft carrier, the *Hōshō*, in December 1919. Following reports and recommendations from several naval officers concerning the state of Japan's naval aviation, the Navy turned to the British Royal Navy, then the most experienced of the world's navies in the field of naval aviation, for assistance. After much discussion, and some reluctance on the part of the British Admiralty, the British Government agreed to send an "unofficial" aviation mission to Japan composed of 30 instructors who had all been

with the Royal Naval Air Service during the war, under the leadership of Colonel the Master of Semphill Sir William Forbes-Sempill.[16] The Semphill Mission brought with it 113 aircraft of various types, ranging from AVRO 504K training planes to Blackburn Swift torpedo planes to Felixstowe F.5 and Supermarine flying boats; later purchases brought the total number of new aircraft to around 200.[17] The aircraft types selected were broadly representative of British naval aircraft (five of the aircraft were then in service with the Royal Navy), and reflected the training needs of the Japanese Navy from basic training to tactical training in reconnaissance with flying boats and observation aircraft, aerial torpedo dropping, gunnery and air combat, and deck landing. The Semphill Mission included several aviation engineers from British aviation manufacturers to provide technical assistance. The Sempill Mission arrived in Japan in the spring of 1921, and selected members of the Mission remained in Japan until early in 1924, enabling Japanese naval aviation to make a "quantum jump in aviation training and technology."[18]

With its first aircraft carrier under construction the Navy needed carrier aircraft. Rather than building foreign types under license the Navy approached the Mitsubishi Aircraft Company, Ltd., which had registered as an airplane manufacturing company in May 1920, to design and build a carrier fighter, reconnaissance plane, and a torpedo plane. Mitsubishi hired an English aeronautical engineer, Herbert Smith, from the Sopwith Aviation Company in early 1921. Smith brought with him to Japan several other English engineers. Working with the Mitsubishi staff, Smith and his team designed the Type 10 Carrier Fighter (the first purpose-built carrier fighter in the world), the Type 10 Carrier Reconnaissance Aircraft, and the Type 10 Carrier Torpedo Aircraft. The Type 10 Fighter and Type 10 Reconnaissance aircraft were successful designs, 128 of the former and 159 of the later type being built between 1921 and 1930; the Type 10 Torpedo aircraft, a triplane design, did not meet the Navy's expectations and only 20 were built. Instead, Smith designed the Type 13 Carrier Attack Aircraft (which was later re-designated the B1M1), which served well into the 1930s, some 442 being built by Mitsubishi and the Navy's Hiro Arsenal. These aircraft became the backbone of the Navy's carrier squadrons during the 1920s.[19]

The 1922 Washington Navy Treaty's limitation on capital ships gave new impetus for the development of the aircraft carrier, especially as the Treaty, with its 5:5:3 ratio of capital ships for the United States, the United Kingdom, and Japan, placed Japan in an inferior position. But it was not until the late 1920s that Japan could do more than experiment with its few carrier airplanes and its sole carrier, the *Hōshō*. As a result of the 1922 Treaty two battle cruisers which the Japanese had already laid down were allowed to be converted into aircraft carriers. These became the *Akagi* and the *Kaga*, and with the formation of the First Carrier Division in 1928 the Navy began to more seriously experiment with the role of carriers in support of the fleet.[20] While work on the *Akagi* and *Kaga* was underway the Navy, in 1926, sought a replacement for its Sparrowhawk and Type 10 Carrier Fighters requesting the Aichi, Mitsubishi, and Nakajima companies to submit proposals. Nakajima decided to approach the Gloster Aircraft Company to acquire the rights for the Gloster Gamecock fighter. Gloster proposed instead the new Gambet, a naval fighter with an air-cooled Bristol Jupiter engine then under development. Nakajima acquired a pattern aircraft, and after some redesign for Japanese production requirements and replacement of the Bristol with a Nakajima licensed-built Jupiter engine, the new fighter won the competition, entering service in 1929 as the Type 3 Carrier Fighter (A1N1).[21] A year later the Navy accepted the Mitsubishi-built Type 89 Carrier Attack Aircraft (B2M1) using a new designation system adopted after the beginning of the Showa era. The Type 89 Carrier Attack Aircraft

came out of a 1928 design competition among Aichi, Kawanishi, Mitsubishi, and Nakajima for a replacement for the Type 13 Carrier Attack Aircraft. Mitsubishi subcontracted its design to Herbert Smith, the Blackburn Aircraft Ltd., and the Handley Page Company, ultimately choosing Blackburn's T.7B design, which featured an all-metal wing structure and a steel-tube fuselage. Although not without problems in operation, the Navy purchased 205 aircraft from Mitsubishi between 1930 and 1935.[22]

From the time it acquired its first aircraft, the Japanese Navy had seen aerial reconnaissance as one of the primary missions of naval aviation. The Navy continued development of both long- and short-range reconnaissance aircraft that could operate from shore bases, seaplane tenders, or be catapulted off the Navy's capital ships. Impressed with the Felixstowe F.5 that the Sempill Mission brought to Japan, the Navy arranged with Short Brothers to build the F.5 under license. A group of Short Brothers engineers travelled to Japan to help the Navy's Hiro and Yokosuka Arsenals begin production; the Aichi Company built 40 additional F.5 flying boats up to 1929.[23] In 1926 the Navy instructed the Hiro Arsenal to design a replacement for the F.5, which was developed as the Navy Type 15 Flying-boat (H1H1), with 20 aircraft built at Navy arsenals and an additional 45 built by the Aichi Company.[24] As a shorter range reconnaissance seaplane in 1922 the Navy adopted the Hansa-Brandenburg W 33, which Japan had received as war reparations. The Aichi Company built 150 of the Navy Type Hansa Reconnaissance Seaplane, and the Nakajima Company built an additional 160 by the end of 1925.[25] The Hansa seaplane served into the late 1920s, when they were replaced by the Type 14 Reconnaissance Seaplane developed at the Yokosuka Naval Arsenal, with 320 being built at the Arsenal and by Aichi and Nakajima, and the Type 15 Reconnaissance Seaplane from Nakajima, which came out of a design competition between the Kawasaki and Mitsubishi companies.[26]

Although the Navy's planned expansion of its air arm had been delayed due to budgetary constraints, by the end of the 1920s the Navy was nearing its goal of having 17 land-based fighter, attack, reconnaissance, and flying boat and seaplane squadrons in addition to the air wings of the First Carrier Division. While smaller than the Japanese Army Air Force in numbers of aircraft, the Navy's aviation arm was more diversified, with more missions and more different types of aircraft than the Army. And, to a greater extent than the Army, the Navy's air arm was beginning to produce a number of capable, air-minded officers who would become strong advocates for the role of naval aviation in the following decade.

The Development of Civil Aviation

The 1920s saw the birth of commercial air transport in Japan. On June 4, 1922, Choichi Inouye established the Japan Air Transport Research Institute (Nihon Kokuyuso Kenkyujo) near Osaka and began an airmail service between Osaka and the island of Shikoku with several seaplanes. A few months later, in cooperation with the Imperial Aeronautical Association and with support from the *Asahi Shimbun* newspaper, Inouye helped set up the East-West Regular Air Transport Association (Tozai Teiki Kokukai), which inaugurated a twice-weekly air mail service between Osaka and Tokyo in January 1923. The third of the early air lines was the Japan Aviation Company Ltd. (Nippon Koku K.K.), organized with the support of the Kawanishi Aircraft Company Ltd. This soon became the Japan Air Lines (Nihon Koku Yuso K.K.) when it was separated from the Kawanishi Company. This company began a service between Osaka and Beppu, on the island of Kyushu, in July 1923; this was later extended to Fukuoka, and then on to Seoul, Korea, and Dairen, China. These three air lines continued to build and extend their services with support from the Government in the form of air mail subsidies

using a combination of imported German aircraft, including the Dornier Komet and Wal and the Junkers F.13, some surplus military aircraft, and domestically designed airplanes from Kawanishi, Mitsubishi, and other Japanese manufacturers. Recognizing the importance of air transport, in 1928 the Japanese government decided to form a semi-governmental air line, and after negotiations with the East-West Regular Air Transport Society and the Japan Air Lines Company, the Japan Air Transport Company (Nihon Kokuyuso K.K.) was organized in October 1928 through an amalgamation of these earlier air lines and the government-sponsored company. Japan Air Transport began service in April 1929 using Army surplus Salmson biplanes for air mail flights between Sendai, Tokyo, Osaka, and Fukuoka. The company purchased six Fokker F.11 Super Universal aircraft from the Atlantic Aviation Company in the United States, a subsidiary of Fokker. When a competition among domestic aircraft manufacturers failed to produce a suitable design, the Nakajima Company acquired the rights to build the Super Universal under license. With these larger aircraft Japan Air Transport began a passenger service within Japan in June 1929, which was extended to Seoul and Dairen in September.[27] That same year the company purchased nine Fokker F.7B/3M aircraft which went into service on the Seoul-Darien route in 1930.

There were a number of notable flights during the period. Numerous European and American pilots attempted record long-distance flights, many stopping off in Japan. The first visit of foreign airplanes took place in May 1920 when two Ansaldo SVA airplanes of the Italian Army Air Service arrived from Rome. Four years later the U.S. Army Air Service's three Douglas World Cruisers completed a trans-Pacific flight from Seattle to Japan via the Aleutians, stopping briefly before continuing their successful world flight. Following several other flights, the air-minded Asahi Shimbun newspaper sponsored a Japanese flight to Europe. The newspaper purchased two Breguet 19 airplanes which were named *Hatsukaze* (First Wind) and *Kochikaze* (East Wind). The airplanes departed Tokyo on July 25, 1925, travelling through Siberia to Moscow, then on to Berlin, Paris, London, and Rome, completing the flight in October. There were several attempts to organize a trans-Pacific flight but without success.[28]

The Japanese aviation industry expanded in both number of firms and capabilities by the end of the decade. The Mitsubishi firm entered the aircraft manufacturing business when it registered its subsidiary, the Mitsubishi Internal Combustion Engine Manufacturing Company (Mitsubishi Nainenki Seizo K.K.), as an airplane manufacturing company. This became the Mitsubishi Aircraft Company (Mitsubishi Kokuki K.K.) in 1928. The somewhat oddly named Aichi Watch and Electric Manufacturing Company (Aichi Tokei Denki K.K., later the Aichi Aircraft Company) also began manufacturing airplanes in 1920, initially for the Japanese Navy. The other major aircraft manufacturers were the Kawanishi Aircraft Company (Kawanishi Kokuki K.K.), which began as the Aeroplane Department of the larger Kawanishi Machinery Manufacturing Works (Kawanishi Kikai Seisakusho) formed after Seibei Kawanishi's break with Chikuhei Nakajima, and the Kawasaki Aircraft Company (Kawasaki Kokuki K.K.), which similarly began as the Aeroplane Section, then Aeroplane Department of the Kawasaki Dockyard Company (Kawasaki Zosensho K.K.). These four companies were the principal airplane and engine manufacturers during the 1920s and 1930s

with the active support and encouragement of the Japanese Army and Navy, who continually urged the companies to increase the domestic content of their airplanes. To speed up the transfer of technology and experience, each of these Japanese firms established close links with European airplane and engine manufacturers, initially in France, but later building relationships with English and German firms and individual aeronautical engineers, inviting several, as previously related, to work in Japan. Mitsubishi, for example, received assistance from Junkers and Rohrbach in Germany and brought over Herbert Smith from Sopwith, while Kawasaki established a relationship with Dornier and hired Dr. Richard Vogt to help with its designs. Several manufacturers started building European airplanes under license as an interim step, then moved to developing their own designs with the help of foreign aeronautical engineers working in Japan under contract, or seeking designs from abroad for modification to Japanese requirements in Japan. In addition, both the Army and Navy imported foreign airplanes for study and sponsored the development of experimental airplanes, all in an attempt to build an independent, indigenous aviation industry. It was an intensive apprenticeship in aeronautical engineering, almost a force-feeding of airplane and engine design and production, but by the end of the decade the Japanese aviation industry was poised for the next jump to true independence and the design of world class airplanes independent of foreign assistance.[29]

Postcards from the Apprenticeship Era

Army Aviation

各務ケ原ニテ

後列前列

杉山航空大長

タレマン少尉
フォル大佐

有馬中将
吉松中将

仁田原大将

ルヘプル少佐
東郷元帥

ベルダン少尉
上原参謀総長

ニッケル少尉
本郷大将

ケルゴレー中尉
藤井大将

岐阜竜泰珍業発行

Fleet Admiral Marquis Heihachirō Tōgō, victor at the Battle of Tsushima, visiting the French Aviation Mission at Gifu in 1919. The Fleet Admiral stands fourth from left in the front row. Lt. Col. Fauré stands second from the right.

A French instructor watches as a pilot trainee and his instructor run up the engine on a Nieuport 83E2 trainer. This type was built under license by the Nakajima Aircraft Company as the Type Ko-2 Trainer.

A French instructor, Vuarin, talks to several Army pilots attending the aerial bombing course at Mikatagahara, in Shizuoka Prefecture.

力馬十二百　ス一ピツツ用逐駆

In 1919 the Japanese Army imported 50 Sopwith Pup fighters as part of its post-war expansion.

（力馬十二百式ロ）機鬪戰型三式甲ルーポーユニ

The Nieuport 24C1, which the Army adopted as the Type Ko 3 Fighter-Trainer. The Nakajima Aeroplane Company built 102 of these airplanes under license.

乙式一型偵察機
サルムソン 230馬力

The Salmson 2A2. As the Type Otsu 1 Reconnaissance Aircraft, the Salmson served with the Army's reconnaissance squadrons into the early 1930s. The Kawasaki Aircraft Company built 300 in addition to 300 built by the Army.

飛行第三聯隊　今飛行セントスル戦闘機ノ雄姿

The Nieuport 29C-1, which was built under license to replace the earlier SPAD 13 and Nieuport fighters. Nakajima built 608 as the Army Type Ko 4 Fighter between 1923 and 1932.

A pair of Type Otsu 1 Reconnaissance airplanes serving with one of the Army's air regiments during the 1920s.

秋季演習参加ノ飛行機

Three Nieuport 29C-1 aircraft in flight over Japan during the 1920s.

二式戦小四号
5152

戦闘用飛行機ニユーポール（ルローン120馬力）

A Nieuport 27 which had been imported with the Nieuport 24.

Part of a series of color postcards showing the Army's airplanes of the 1920s; this card shows a SPAD 13. The Army purchased 100 SPADs in 1919.

ス式\型
7403

戦闘用飛行機スパット（イスパノスイザ220馬力）

戦誼用飛行機スパットエルブモン（イスパノスイザ300馬力）

In October 1921, the Army purchased a single SPAD-Herbemont 20C-2, the Army's first two-seat fighter.

In 1922, the Army's Aviation School at Tokorozawa completed an experimental fighter plane incorporating the features of a number of the French airplane types the Army had imported after the War. The Koshiki-2 Experimental Fighter showed promise, but was not developed further.

絞式二型單座驅逐機(サ式230馬力)

A postcard from the same color series showing the NiD-29C-1/Type Ko 4 Fighter.

二式二九型飛行機(ニューポートラザウラージュ別型式)(ナ式300馬力)

The Army purchased one Morane-Saulnier MoS 30 in 1922 for evaluation as an advanced trainer, but it was not adopted.

モラーヌ S.F.A 30型練習機(ルローン 80馬力)

The Farman F.60 *Goliath*, with the Farman F.50, served as the Army's first heavy bombers. Both types participated in the development of both day and night bombing during the early 1920s.

機 爆 型 二 式 T

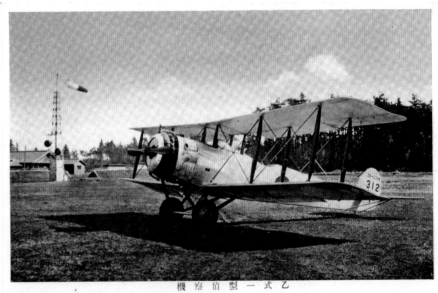

A Type Otsu 1 Reconnaissance airplane on an Army airfield during the 1920s.

機 察 偵 型 一 式 乙

The Army imported a single Hanriot HD-14 two-seat trainer in 1923. Mitsubishi Aircraft acquired the license and built 145 HD-14s for the Army as the Type Ki 1 Trainer to replace the earlier Ko 1 Trainers. The Hanriots remained in service until 1935.

Two Army pilots next to a
Type Ko 3 (Nieuport 24) Fighter.

The Army purchased the
Caudron G 6 twin-engine
trainer in 1921 to train its
bomber pilots as the Type Bo 1.
It served until 1923.

A postcard from
October 2, 1926,
the last year of the
Taisho era, published
for an air day to
promote interest in
aviation. The card is a
mixture of a photo of
a parachutist jumping
out of a Type Otsu
1 Reconnaissance
airplane with
a backdrop of
more traditional
Japanese motifs.

In 1924, the Army placed an order with the Kawasaki Aircraft Company for a new all-metal bomber to replace the older Farman F.50 and F.60. Built with the assistance of Dr. Richard Vogt of Dornier and resembling the Dornier Wal flying boat, the Army adopted Kawasaki's aircraft (the Kawasaki-Dornier N) as the Type 87 Heavy Bomber.

八七式重爆撃機
B. M. W. 400馬力 2個

The Mitsubishi Type Washi (Eagle) experimental light bomber. Mitsubishi Aircraft submitted this design in response to the Army's 1925 competition for a new light bomber.

三菱式輕爆撃機

When its Type Washi failed to win the light bomber competition, Mitsubishi offered the Army a version of its Navy Type 13 Carrier Attack Aircraft, which the Army accepted as the Type 87 Light Bomber. Mitsubishi built 48 of these aircraft between 1926 and 1929.

八七式輕爆撃機

The Army requested designs for a
new reconnaissance airplane in 1926.
Kawasaki's design was the winning
entry, and was adopted as the Type 88
Reconnaissance Aircraft, the first aircraft
with an all-metal structure built in Japan.
This postcard shows the early version
(later designated the Type 88-1) with a
blunt nose.

The Type 88-2 Reconnaissance
Aircraft featured a redesigned
streamlined nose and the addition
of ailerons to the lower wing.
Built in large numbers to replace
the Type Otsu Reconnaissance
Aircraft, the Type 88 served until
1940 and saw extensive combat
in China during the 1930s.

Type 88 Reconnaissance Aircraft of the 4ᵗʰ Flying Regiment (4ᵗʰ Hiko Rentai) flying over the Japanese countryside.

八八式偵察機

The wood and fabric Type 87 Light Bomber was nearing obsolescence by the end of the 1920s. The Army decided to adapt the excellent Kawasaki Type 88 as a light bomber, ordering a total of 407 Type 88 Light Bombers from Kawasaki and the Tachikawa Aeroplane Company. The Type 88 Light Bomber saw active service in the Manchurian Incident in 1931.

A Curtiss P-1A Hawk with the Curtiss D-12 engine that the Army purchased for study purposes in 1927. Interestingly, the postcard says that this airplane was built by Kawasaki. Both the Army and Navy regularly imported foreign airplanes to study developments in aeronautical engineering, as did many other air forces during the period. The Army purchased a later P-6 Hawk with a Conqueror engine in 1930.

川崎製戦闘機

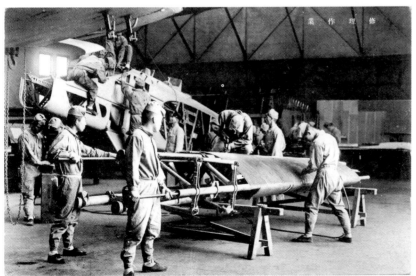

Workers at the Kawasaki Aircraft Company building a Type 88 Reconnaissance aircraft.

Naval Aviation

力馬十四百二 一サンパルナバ

A member of the Sempill Mission stands with several Imperial Japanese Navy officers and enlisted men next to a Parnell Panther. The Panther, a two-seat fleet spotting and reconnaissance aircraft, was then in service with the Fleet Air Arm. The Sempill Mission brought 12 Panthers with them to Japan.

The Sempill Mission brought 20 Avro 504K trainers and 10 Avro 504L seaplane trainers to provide refresher training to Japanese Navy pilots. The Navy adopted the Avro as its primary trainer and obtained manufacturing rights from the A. V. Roe & Company. Nakajima and Aichi built 280 wheel and float versions from 1922 to 1924.

力馬十二百 式四ブブ

力馬十五百式ロブア 機行飛上水

A Navy Avro 504L seaplane trainer, sometimes referred to as the Avro 504S.

Among the ten aircraft types the Sempill Mission brought to Japan was the Felixstowe F.5, built by Short Brothers. The Navy decided to build these airplanes in Japan and engaged a team of engineers from Short to assist in production. Navy arsenals built around 20 F.5s, and the Aichi Company built an additional 40.

行聲山富 面測力馬十五百七號五機フエ製行飛 （製行飛軍海清が慶）

力馬250 ルネンヤチ 機行飛上水

The Sempill Mission brought several smaller flying boats to Japan. This is the Supermarine Channel, which the Mission used for preliminary training in flying boats.

The four-seat Vickers. Viking was another flying boat the Sempill Mission used for training Navy flying boat pilots.

水陸兩用ハ イヤシケン號　（霞ヶ浦飛行場）

The Navy purchased the Supermarine Seal in 1922, apparently for research purposes, as it was not adopted.

水陸兩用 シ ー ル號　（霞ヶ浦飛行場）

魚雷用飛行機　タンク 二百四十馬力

The Sempill Mission also brought six Sopwith Cuckoo torpedo planes to Japan to complement the Blackburn Swifts.

The Blackburn Aeroplane & Motor Company supplied two Blackburn Swift Mk. II airplanes to the Sempill Mission to train Japanese Navy pilots in torpedo-dropping and in simulated deck landing.

スウイフト號　（霞ヶ浦飛行場）

一三式艦上攻撃機から魚雷發射

The second
of the two
Blackburn Swifts
(the numeral 2
can just be made
out ahead of
the horizontal
tail) dropping a
practice torpedo
in 1925.

The Gloster Sparrowhawk, shown here at the Navy's Kasumigaura airfield, was the Navy's first carrier-borne fighter. The Navy purchased 50 Sparrowhawks from the Gloster Aircraft Company, including 10 Sparrowhawk III aircraft equipped with arrester hooks on the landing gear axel.

Senior Naval officers visiting the operations of the Sempill Mission at Kasumigaura airfield. Note the machine guns attached to tripods in the canvas hangar.

Servicing Blackburn Swift No. 1 at Kasumigaura airfield.

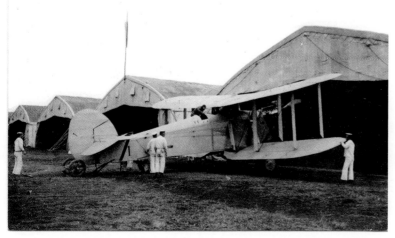

力馬百三 式年十用察偵

Herbert Smith of the Sopwith Company supervised the design of three naval aircraft for the Mitsubishi Aircraft Company. The Navy Type 10 Carrier Fighter was the world's first fighter plane designed for operations from an aircraft carrier. The Type 10-1, shown here, had a honeycomb radiator.

KASUMIGAURA KOKUTAI
機上陸

A row of Navy Type 10-1 Carrier Fighters in the early 1920s. The postcard carries the caption that these aircraft were from the Kasumigaura Kokutai (Air Group), but the tail code R-214 is more closely associated with the Yokosuka Kokutai.

三菱型戰闘機

The Navy Type 10-2 Carrier Fighter had the honeycomb radiator replaced with two Lamblin radiators under the fuselage. This model proved to be some 15 mph slower than the Type 10-1.

機行飛用察偵式年十 （場行飛浦ヶ霞）

Herbert Smith also designed the Navy Type 10 Carrier Reconnaissance Aircraft, in effect a scaled-up version of the Type 10 Carrier Fighter. The Type 10-1, seen in this postcard, had the honeycomb radiator in front of the engine.

A later version of the Navy Type 10 Carrier Reconnaissance Aircraft, with Lamblin radiators under the wings replacing the honeycomb radiator ahead of the engine and the pilot's cockpit moved forward. The aircraft in this view, however, retains the tail and rudder of the earlier Type 10-1.

三菱型偵察機．

海軍飛行機陸上の活躍 （大正15.7.9横須賀撮影） 畠山發行

A section of Navy Type 10 Carrier Reconnaissance Aircraft having just landed on a Navy airfield in 1926.

一三式艦上戦闘機

A postcard showing a Navy Type 10-2 Carrier Reconnaissance Aircraft in flight.

力馬十五百四 式年十葉三菱三 機行飛用雷魚

Herbert Smith's third design for Mitsubishi, the Navy Type 10 Carrier Torpedo Aircraft, was less successful, and only twenty were built.

望遠の艦母空航と機察偵上艦　（機空航軍海）

To replace the Navy Type 10 Carrier Torpedo Aircraft, Herbert Smith developed a more capable aircraft which was accepted as the Navy Type 13 Carrier Attack Aircraft in 1923. This tinted postcard shows a Type 13 Carrier Attack Aircraft in flight with the aircraft carrier *Kaga* or *Akagi* in the background. The Katakana symbol on the tail identifies this airplane as belonging to the Yokosuka Kokutai.

隊艦力主ミ機闘戦上艦　（機空航軍海）

A tinted postcard showing a Navy Type 10 Carrier Reconnaissance Aircraft. The letter A on the tail likely indicates an aircraft from the aircraft carrier *Hōshō's* air group.

機察偵上水
力馬○○二一ザンハ

Land-based seaplanes continued to form an important component of the Navy's air arm. The Navy received several Hansa-Brandenburg W33 airplanes as war reparations and decided to build the type in Japan, placing orders with Nakajima and Aichi, who together built 310 as the Navy Type Hansa Reconnaissance Seaplane.

水上偵察機
ブレゲー二四〇馬力

In 1925 the Navy sought designs for a new long-range reconnaissance seaplane. Nakajima submitted a twin-float design based on the French Breguet 19A2.

Kawasaki submitted a conversion of the Dornier Komet C-3 land transport plane for the 1925 reconnaissance seaplane competition, but it failed to meet the Navy's range requirement.

KASUMIGAURA KOUKUUTAI
水　上　機

行飛乗同ノ年學一

The winner of the long-range seaplane competition was a design from the Navy's arsenal at Yokosuka, which was adopted as the Navy Type 14 Reconnaissance Seaplane (E1Y1), shown here with a group of pilots and observers from the Yokosuka Kokutai.

水上練習機
一三式一三〇馬力

The Navy's engineers at Yokosuka also designed a replacement for the Avro 504K Land-based Trainer and the 504L Seaplane Trainer as the Navy Type 13 Trainer (K1Y1-2). This postcard shows two K1Y2 airplanes belonging to the Kasumigaura Kokutai.

陸上練習機
一三式一三〇馬力

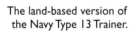

The land-based version of the Navy Type 13 Trainer.

An atmospheric image of a Hansa-Brandenburg W33 alight on a fog-shrouded sea, with the iconic Mount Fuji rising in the distance. What appears to be a Felixstowe F.5 flies above on the right.

攻擊機ノ魚雷發射

An image of a Navy Type 13 Carrier Attack Aircraft dropping a torpedo, with capital ships in the background.

Civil Aviation

In 1926 the *Asahi Shimbun* placed an order with the Kawasaki Aircraft Company for three Kawasaki-Dornier Komet passenger transports for use by the East-West Regular Air Transport Association (Tozai Teiki Kokukai).

三菱式ユンカースF13型旅客輸送機

The Mitsubishi Company purchased two Junkers F.13 transports in 1928. The Japan Aviation Company later put these airplanes into service as seaplanes.

圖山川に着水せる城崎第一號機　（城崎名所）
The Air-chip on the Maruyama river, Kinosaki

In response to a competition initiated by the Aviation Bureau of the Japanese Department of Commerce for improved passenger planes, the Mitsubishi Aircraft Company designed the Mitsubishi MC-1, employing the wings of the Type 13 Carrier Attack Aircraft and a Mitsubishi-built Jaguar engine. The MC-1 was used briefly on the East-West Regular Air Transport Association's Tokyo-Osaka service in 1928-29 in both land-plane and float-plane versions.

In 1929 the newly-formed
Japan Air Transport Company
inaugurated service with six
Fokker F.11 Super Universals
purchased in the United States.
Nakajima Aircraft Company later
acquired the license to build the
F.11, completing 47 airplanes.

One of Japan Air Transport's
Fokker F.7B/3M airliners
at Haneda Airport.

A Fokker F.11 taxiing
out for take-off from
Toyama, a city located
on the west coast
of the main island of
Honshu.

氏　章藤佐　ゝ　號　ラキア

Tomotari Inagaki designed the Itoh 19 *Akira-Go* for Akira Sato (formerly Yozo Sato), who used the airplane for barnstorming in parts of Japan that had not seen an airplane.

On May 21, 1920, two Ansaldo SVA airplanes arrived in Tokyo after a 9,476 mile flight from Rome, the first visit of a foreign airplane on a flight to Japan.

（日伊大飛行）　マ　中西西の通の世　る飛行関

In April 1921 Barr's Flying Circus came to Japan for a series of exhibition flights, but by this time airplanes were no longer a novelty.

況實ノ行飛大リ乗曲行一氏一バ　　（家行險冒國米）

尉中三ンソルネ・ドーエウ・スもりよ左　（念記來飛行機一周世界訪來記念）

Lieutenants Lowell Smith, Leigh Wade, and Erik Nelson, members of the American Army Air Service's First World Flight, after landing in Japan in May 1924.

ぎかは便郵

九三式双軽爆撃機

The Years of Independence

1931 – 1940

The Thirties:
The Rise of Militarism

THE DECADE OF THE 1930s WAS A TURBULENT time for Japan. Politically the nation began the decade as a parliamentary democracy and ended it under a de facto military dictatorship, with the military exercising a dominant role in governing the nation, controlling the direction of the economy, and determining the course of Japan's foreign relations. The course of events in the proceeding decade– the intense competition among political parties, the rise of big business and the expansion of industrialization, the continued growth in the urban population, the adoption of Western cultural attributes among some segments of the society, and the emergence of more radical socialist and communist ideologies–created within certain segments of the society, especially within the military, a deep-seated sense of dissatisfaction with Japan's apparent path and a desire to return to more traditional Japanese values.[1] Some saw the establishment of the Soviet Union and a resurgent and nationalistic China as direct threats to Japan's position on the mainland of Asia, and thus to the nation's future. A viewpoint emerged among a number of younger Army officers that Japan was headed toward an inevitable war with the Western powers, and to preserve the Japanese emperor, people, and culture the nation had to be mobilized toward total war, utilizing the resources of the rest of Asia for Japan's own preservation and the expulsion of the Western colonial powers.[2] The means to this end, to ensure Japan's very survival, required the Army's intervention in all aspects of Japanese life. From the late 1920s on the Army, which was the more overtly political of the two services, intervened directly in the political sphere and initiated a program of aggressive expansion in Manchuria and China independent of civilian oversight or control. The onset of the Great Depression and the seeming failure of both capitalism and liberal democracy saw the rise in Japan of "an increasingly strident and aggressive nationalism", characterized by anti-foreign attitudes, the championing of Japan's leading role in Asia, and a drive for economic autarky.[3]

In September 1931, the Japanese Kwantung Army in Manchuria, on its own initiative but with the tacit support of the Imperial Army Headquarters in Japan, orchestrated an incident that led to the occupation of much of Manchuria and the establishment of a nominally independent state of Manchuko. After a critical report from the League of Nations Japan left the League and abandoned internationalism to follow its own path; a short time later Japan rejected the naval limitations imposed in the 1922 Washington Naval Treaty and reaffirmed in the London Naval Treaty of 1930. As the decade went on the Army and the Navy embarked on a program of rearmament; between the fiscal years 1930-31 and 1936-37, the combined budget for the Army and Navy more than doubled.[4] The Army grew from 17 divisions in 1931 to 50 divisions in 1940, while the Navy's total tonnage grew from 850,000 tons to over 1 million tons during the same period.[5] Following the outbreak of war with China in July 1937, the defense budget regularly absorbed 60% or more of total government expenditures. Even with the vast expansion in armament, however, the Army and Navy's strategic goals did not always align with the military's operational capabilities or with Japan's industrial capacity.[6] The Army embarked on a campaign in China expecting a rapid victory

over a demonstrably weaker opponent, only to find that Chinese resistance could not be easily broken. Despite substantial territorial conquests, the capture of the main Chinese port cities, and an attempt at economic strangulation, by 1941 Japan was entangled in a war that "...could not be won, either by military or other means, could not be reduced to manageable proportions, and that was prohibitively expensive."[7]

During the 1930s the Japanese aviation industry made impressive strides in design and production. Both the Army and the Navy were intent on establishing an aviation industry independent of foreign technological support.[8] In April 1932 the military adopted the "Independent Aircraft Technology Plan" (Kōkū Gijutsu Jiritsu Keikaku), which specified that only Japanese engineers could design airplanes and engines for the Japanese military.[9] With the impetus of military rearmament and increasing budgets the aviation industry flourished. During the decade of the 1930s the industry produced almost 5,000 airplanes, considerably more than in the prior decade.[10] More importantly, the aviation industry made the technological transition to the all-metal monoplane and achieved a higher degree of independence from foreign manufacturers, although never completely freeing itself from foreign technology, particularly in aircraft engines. But by the end of the decade the primary combat aircraft for both the Army and the Navy, with few exceptions, came from indigenous designs. In several categories Japanese aeronautical engineers developed airplanes that were not simply comparable to their Western counterparts, but truly exceptional for their time. Evidence of the Japanese aviation industry's rapid progress came through several record long-distance flights and Japanese air operations over China, which surprised Western observers.

The Development of Army Aviation

At the beginning of the Manchurian Incident in September 1931, the Army sent out a squadron of Type Ko 4 Fighters (Nieuport 29) and a squadron of Type Otsu 1 Reconnaissance aircraft (Salmson 2A2) to support the Army's advance, soon adding additional squadrons of Type 88 Reconnaissance aircraft and Type 87 Heavy and Light bombers. Facing next to no aerial opposition from Chinese forces, the six squadrons sent to Manchuria spent the next several months in almost continuous operations in support of the Army's advance. Toward the end of the conflict the newer Nakajima Type 91 Fighter began to replace the now elderly Type Ko 4, while the Mitsubishi Type 92 Reconnaissance Aircraft supplemented the Type 88 aircraft and older Type Otsu 1. Army aircraft also became involved in the fighting around Shanghai in early 1932 supporting the Japanese Navy's air units. For the next fourteen years Japanese Army air units would maintain a continuous presence in Manchuria and China.[11]

The Manchurian Incident saw the beginning of donations of aircraft to the Army purchased with funds raised through popular subscription among the general public, civic and patriotic associations, businesses, municipalities, and prefectures. These donated airplanes were dubbed "Aikoku", meaning love of country. The Army Ministry would typically paint the name of the donating organization on the fuselage of the airplane and specially number the aircraft. The aircraft were dedicated at a special ceremony where Shinto priests would formally name the aircraft, with senior Army officers and members of the donating organization in attendance. The Army Ministry prepared special packets of postcards to commemorate the event, which would include a photograph or a painting representing the specific aircraft placed in an envelope with the name of the donors on the cover. The first donation of an aircraft took place in January 1932, resulting from a campaign organized with the support of the *Asahi Shimbun* and *Mainichi Daily* newspapers. Funds raised through this campaign went to purchase a Junkers K-37 bomber which the Mitsubishi Aircraft

Company had imported for research purposes, and a Dornier Do-B Merkur that the Kawasaki Aircraft Company had imported and converted to an ambulance aircraft. The Junkers K-37 became *Aikoku 1* and the Dornier Do-B *Aikoku 2*. The two aircraft were given to the Army in a special ceremony in Tokyo and then flown to Manchuria. These donations proved quite popular, with the Army receiving some 54 airplanes over the next year. The practice continued almost up to the end of World War II, by which time the Army had probably received over 6,000 aircraft through patriotic donations.[12]

The success of the Kwantung Army's Manchurian adventure and the establishment of Manchuko brought new challenges and greater responsibilities for the Japanese Army Air Force, as well as the benefits of actual combat experience. Japan's control over large areas of Manchuria brought Japanese Army forces closer to the border with Russia and within range of Russian long-range bombers. The Army decided to acquire more capable bombers to replace the elderly Type 87 Heavy bomber, which had proved inadequate in operations, and improved fighters to ensure that the Army Air Force could maintain air superiority.[13] Impressed with the performance of the Junkers K-37, the Army turned to Mitsubishi Aircraft to design a new heavy and new light bomber along the same lines. Mitsubishi developed the Type 93 Heavy Bomber (Ki-1) and the Type 93 Twin-engine Light Bomber (Ki-2), both of which lasted in service to the early stages of the Sino-Japanese War.[14] At the same time the Army accepted the Type 93 Single-engine Light Bomber (Ki-3) from Kawasaki to replace the Type 88 Light Bomber and the Kawasaki Type 92 fighter to supplement the Type 91 Fighter.[15] From Nakajima came the Type 94 Reconnaissance aircraft to replace the Type 88 aircraft. These aircraft were transitional, more capable but still incorporating a bi-plane configuration, or in the case of the early Ki-1 and Ki-2 bombers, fixed landing gear. Following the conclusion of the Manchurian conflict, the

Army embarked on an expansion of its air units in Japan and in Manchuria, initially to a total of 36 squadrons, and then to 53 squadrons by 1936, at which time the Army had 21 fighter squadrons, 12 reconnaissance squadrons, 12 light bomber squadrons, and 8 heavy bomber squadrons.[16] To control the new air regiments the Army created a new command, the Air Group (Hiko-dan), under the overall control of an Air Command (Koku Heidan). Even before completion of this expansion, the Army General Staff made plans to add an additional 89 squadrons over a six-year period, to be completed by 1942.[17]

The middle years of the decade saw the Japanese Army Air Force initiate a modernization program that led to the development of the aircraft types that would see the Air Force through the Sino-Japanese war and into the early years of World War II. Beginning in 1935 the Army initiated competitions for a new fighter, reconnaissance aircraft, heavy bomber, and light bomber. The first of this new generation to emerge was the Mitsubishi Type 97 Command Reconnaissance Aircraft (Ki-15), which Mitsubishi Aircraft designed to meet an Army requirement for a fast reconnaissance aircraft. For its day the Ki-15 had an exceptional performance for a reconnaissance aircraft, with a top speed of 299 mph. The Army requested Mitsubishi and Nakajima to design a twin-engine heavy bomber to replace the Mitsubishi Ki-1. Mitsubishi's entry won the Army's approval and entered service as the Type 97 Heavy Bomber (Ki-21), with production split between Mitsubishi and Nakajima. For its new light bomber the Army turned to Mitsubishi again, and to Kawasaki. Mitsubishi designed the radial engine Type 97 Light Bomber (Ki-30), while Kawasaki chose a liquid-cooled inline engine for its Type 98 Light Bomber (Ki-32), and the Army adopted both types. In 1934 the Army had considered some monoplane fighter designs from Kawasaki and Nakajima but chose a more conservative approach, accepting a traditional biplane design from Kawasaki as the Type 95 Fighter (Ki-10). Recognizing that the

biplane was nearing the limit of its capability, in early 1936 the Army initiated a new competition for a more advanced monoplane fighter, soliciting designs from Kawasaki, Mitsubishi, and Nakajima. Kawasaki's entry, the Ki-28, was faster and had a superior rate of climb to the other two entries, but with the Army pilots' obsession with superior maneuverability, the supremely agile Nakajima Ki-27 carried the day, entering service as the Type 97 Fighter.[18] In 1937 the Army issued a requirement for a new Army co-operation airplane, choosing a Tachikawa Aeroplane Company's design as the Type 98 Direct Co-operation Airplane (Ki-36).

The Sino-Japanese War began as the Army Air Force's modernization program was getting underway, so the Army's air units went to war with the older airplane types from the early and mid-1930s. During July 1937 the Army mobilized 24 squadrons from Japan, Manchuria, Korea, and Formosa, and rushed them to the scene of fighting in northern China.[19] These squadrons were equipped with the Type 93 Heavy and Light bombers, Type 94 Reconnaissance aircraft, and the Type 95 Fighter. These aircraft supported the Army's advance through northern and central China over the next year, providing close support to the infantry, reconnaissance, bombing Chinese positions, and maintaining air superiority over the battlefield. The Mitsubishi Type 97 Command Reconnaissance aircraft was the first of the modern airplanes to enter combat, proving faster than almost all Chinese Air Force fighters available at the time. The Mitsubishi Type 97 Heavy Bomber entered combat in November 1937, followed by the Nakajima Type 97 Fighter, which made its combat debut in April 1938, and the Mitsubishi Type 97 Light Bomber, which began operations in China a few months later. Army Air Force units in China were progressively re-equipped with the new generation of aircraft over the succeeding months. During the summer of 1938, the Army Air Force underwent a thorough reorganization and a new type of unit, the *Hiko Sentai*, was created to replace the *Hiko Rentai*, removing some of the administrative functions and grouping similar types of aircraft into a single homogeneous unit comprised of three or more squadrons.[20] In the midst of the China conflict the Japanese Army became involved in an undeclared war with the Soviet Union in Mongolia. In the Nomonhan Incident, lasting from May to September 1939, the Japanese Army, for the first time, came up against a modern mechanized Army in combined arms warfare and was roundly defeated. The Japanese Army Air Force was heavily involved in the fighting, with large formations of bombers and fighters clashing over the Mongolian plains. In the opening phases of the conflict the experienced Japanese fighter pilots made short work of their less experienced Russian counterparts, but when the Russian Air Force brought in more experienced pilots and changed tactics to avoid individual dogfights, the Japanese were more hard-pressed against the more heavily armed Polikarpov I-153 and I-16 fighters using hit-and-run attacks. Still, the Japanese fighter Sentais claimed to have shot down 1,162 Russian aircraft, six times the number that were actually lost.[21]

Despite requests from pilots for heavier fighters with more firepower, the Army Air Force's experience in China and the apparent successes during the Nomonhon Incident bred a certain complacency in Army air circles and failed to dislodge the obsession with maneuverability. When the Army requested Nakajima to develop a replacement for the Type 97 Fighter maneuverability remained the paramount requirement, delaying introduction of what would become the Nakajima Type 1 Fighter (Ki-43) until just before the outbreak of the Greater East Asia War. The Army began development of heavy fighters belatedly, and these would not emerge until the middle years of Pacific War, and would never be built in the quantities needed. Several other airplanes emerged in the last years of the 1930s that would see service in the great battles just ahead.

As a replacement for the Mitsubishi Type 97 Heavy Bomber, in March 1941 the Army accepted a Nakajima design as the Type 100 Heavy Bomber (Ki-49). The Army turned to Mitsubishi to develop a replacement for its Type 97 Light Bomber, and Mitsubishi developed the Type 99 Assault Plane (Ki-51). At the same time, Mitsubishi developed a replacement for its Type 97 Command Reconnaissance Airplane, the twin-engine Type 100 Command Reconnaissance Airplane which, when it was introduced in 1940-41, was faster than the best Army and Navy fighters. For the first few years of the Pacific War the Type 100 would be nearly impervious to interception. On the eve of the Greater East Asia War the Japanese Army Air Force had a total of 150 squadrons (mainly 55 fighter, 33 light bomber, 32 heavy bomber, and 29 reconnaissance) with 1,570 airplanes.[22] The Army Air Force now had four years of combat experience in China and against Russia in the Nomonhan, a well-developed tactical doctrine, and a force of airplanes that, on the surface, seemed in the Army's eyes to be more than adequate for the task ahead. What the Japanese Army Air Force had failed to appreciate from the experiences of the combatants in the larger air war in Europe was that, in the coming fight with the Allied powers speed, firepower, defensive armament, armor protection, and weight of bombs would be paramount, all features that were lacking in the Army's principal aircraft.[23]

The Development of Naval Aviation

In January 1932 fighting broke out in Shanghai between Japanese Imperial Navy Marines and local Chinese forces, resulting in some fierce fighting in and around Shanghai. The Navy command in Shanghai quickly called for air support. The seaplane tender *Notoro* arrived first and launched attacks with its Type 14 and Type 90 Reconnaissance Seaplanes. The carriers *Hōshō* and *Kaga* arrived a few days later with their air groups equipped with the Mitsubishi

Navy Type 13 Carrier Attack bombers (B1M3) and Nakajima Navy Type 3 Carrier Fighters (A1N2). The fighting around Shanghai gave the Imperial Japanese Navy its first experience of combat operations with carrier-borne aircraft. More notably, in the brief fighting around Shanghai the Navy scored Japan's first aerial victory when, on February 22nd, the American pilot Robert Short took off in a recently imported Boeing Model 218 (P-12) to intercept a flight of Type 13 bombers from the *Kaga*. Short attacked the bombers and fatally wounded the lead pilot, Lt. Susumu Kotani, but was then attacked in turn by the escorting flight of Type 3 fighters led by Lt. Nogiji Ikuta, who managed to shoot down the Boeing fighter, killing Short. A few days later five Chinese Air Force airplanes attacked a formation of Japanese bombers and their escorts, with the Type 3 fighters claiming three Chinese aircraft shot down.[24]

In the aftermath of the Shanghai fighting the Imperial Navy embarked on a program intended to make the Japanese aviation industry self-sufficient and reduce its dependence on foreign technological support. Rear Admiral Isoroku Yamamoto, then head of the Technical Bureau in the Navy's Aviation Department, had been the driving force behind the "Independent Aircraft Technology Plan" referenced above. At the same time the Navy established an aeronautical development center at the Yokosuka Navy Air Arsenal by bringing its various technical research efforts together. Thereafter the Navy Air Arsenal coordinated all technological development related to aviation. At the same time, the Navy Air Arsenal instituted a system of managed competition designed to support the domestic aviation industry. The new system called for two companies to compete in the design of aircraft for the Navy, but the winning design would also be placed in production with the losing manufacturer, thereby spreading the newest and best technology and enabling more firms to sustain their business. Over the next several years the Japanese Navy would stimulate

the development of several exceptional naval aircraft.[25]

The Imperial Navy decided, as the Army had done, to expand its aviation arm. An added impetus was the continued restrictions on the number of capital ships allowed Japan as agreed in the 1930 London Naval Treaty. During the middle years of the decade the Navy evaluated some 50 aircraft types as its requirements and aviation technology underwent considerable change.[26] The Navy sought to increase both its carrier-based and land-based aviation as a means of compensating for its inferiority in capital ships, still deemed the primary element of the fleet. One of the first competitions was for a new carrier attack airplane to replace the older Type 13 and Type 89 models. When designs from Aichi, Mitsubishi, and Nakajima proved to be unacceptable, the Navy turned to the Yokosuka arsenal for the Navy Type 92 Carrier Attack Aircraft (B3Y1), and when engine problems developed in this airplane, a second competition led to the adoption of the Yokosuka Navy Type 96 Carrier Attack Aircraft (B4Y1), which was the Navy's last fabric-covered carrier bomber.[27] A competition for a new carrier fighter saw the development of the Nakajima Navy Type 95 Carrier Fighter (A4N1), the Navy's last biplane fighter, which was only a modest improvement over the older Type 90 Carrier Fighter it was meant to replace.[28] A number of these aircraft came to the Navy through patriotic donations in a program similar to the Army's. Many different organizations raised funds to purchase aircraft for the Navy. These were known as *Hokoku* aircraft, meaning "service to one's country". The practice continued through World War II. As part of the Navy's expansion program, the Navy established new land-based squadrons and commissioned two new carriers, the *Hiryu* and the *Soryu*.

New aircraft types that the Navy began to develop during these years were the dive bomber and the long-range, land-based bomber. Both types grew out of new thinking about the role of both carrier and land-based aviation in the event of a conflict with the U.S. Navy, the Japanese Navy's most likely adversary. Realizing that it would begin any naval conflict from a position of inferiority in capital ships, the Japanese Navy sought ways in which it could reduce the strength of the U.S. Navy's fleet through attrition as it sailed across the Pacific before what was expected to be a final confrontation. Following the progress of dive bomber development in the U.S. Navy in the early 1930s, the Japanese Navy came to the realization, as had the U.S. Navy, that dive bombers could weaken, if not sink, a capital ship, and could also be used to damage an opposing force's aircraft carriers.[29] In 1933 the Navy issued a requirement for a carrier dive bomber. Aichi submitted a modified Heinkel He 66 dive bomber obtained through a licensing agreement with Heinkel and won the competition, receiving a contract to build the type as the Navy Type 94 Carrier Bomber (D1A1), the Navy's first dive bomber, later superseded by the improved Navy Type 96 Carrier Bomber (D1A2) built in larger numbers.[30] Given the relatively short range of carrier aircraft at the time, an argument emerged within Navy aviation circles that what Japan needed was a land-based aircraft with exceptional range to strike the U.S. fleet well before it neared Japan, using the mandated islands as forward bases.[31] Rear Admiral Yamamoto advocated the same idea, and as head of the Technical Bureau supported development of two experimental designs, the Hiro Naval Arsenal's Type 95 Land-Based Attack Aircraft (G2H1) and the Mitsubishi Experimental 8-Shi Special Reconnaissance Aircraft (G1M1), which had the remarkable range of 2,280 nautical miles.[32] While neither of these aircraft were a complete success, they demonstrated that the concept had merit and led the Navy to place an order with Mitsubishi Aircraft for a long-range land-based bomber based on the G1M1. The prototype flew in July 1935 and was adopted the following year as the Type 96 Land-Based Attack Aircraft (G3M1). The G3M1's operational debut in the

first few months of the Sino-Japanese War would astonish western observers.[33]

In the middle years of the decade the Navy began the transition from fabric-covered biplanes to all-metal monoplanes. In 1934 the Navy issued a request for a new carrier fighter. Building on its experience with an earlier monoplane design, Mitsubishi's designer Jiro Horikoshi developed a low-wing, fixed-gear monoplane which exceeded the Navy's already stringent requirements for speed and rate of climb, and was perhaps the best carrier fighter in the world at that time.[34] The new fighter entered service in early 1937 as the Type 96 Carrier Fighter (A5M1). A year after issuing the request for a new fighter, the Navy set out a requirement for a new all-metal, monoplane torpedo bomber. Both Mitsubishi and Nakajima submitted designs, Mitsubishi choosing a fixed landing gear for its B5M1 while Nakajima's entry, the B5N1, had retractable landing gear; both aircraft had folding wings for easier storage on a carrier. The Navy accepted both aircraft as the Type 97 Carrier Attack Bomber, but subsequently ordered the Nakajima B5N in greater quantity, as it proved more successful in service.[35] With a fighter and torpedo bomber in hand the Navy then sought a new monoplane dive bomber to replace the older Type 96 Carrier Bomber (Aichi D1A1-2). A specification issued in the summer of 1936 brought three design proposals from Aichi, Mitsubishi, and Nakajima, with Aichi winning the competition with its D3A1, which went into service as the Type 99 Carrier Bomber.[36] Other notable airplanes that emerged during this period and saw service to the end of the decade were the Kawanishi Type 94 Reconnaissance Seaplane (E7K) and the Nakajima Type 95 Reconnaissance Seaplane (E8N), both single-engine float planes used aboard seaplane tenders, battleships, and cruisers, and the Kawanishi Type 97 Flying Boat (H6K) which, in its later versions, had a maximum range of 3,600 miles, greater than the comparable Short Sunderland and Consolidated PBY flying boats.

The Imperial Navy was quickly drawn into the fighting in China following the outbreak of hostilities on July 7, 1937. Under an agreement between the Army and the Navy general staffs, the Navy was assigned responsibility for operations in central and southern China. Within a few weeks the 1st Air Wing (1st *Koku Sentai*), with the aircraft carriers *Ryujo* and *Hōshō*, and the 2nd Air Wing (2nd *Koku Sentai*) with the *Kaga* were patrolling off Shanghai.[37] Like the Army, the Navy was in the midst of its modernization program, so certain of the Navy's older aircraft had to carry the burden of its operations during the early months of the conflict, flying alongside the newly introduced Type 96 Carrier Fighter (A5M) and the Type 96 Land-based Attack Bomber (G3M). The carrier-based aircraft were soon involved in attacking Chinese air bases and providing support to the ground forces fighting around Shanghai. On August 14 and 15, 1937, the First Combined Air Group, composed of the Kisarazu and Kanoya Air Groups with 42 G3M bombers, flew from western Kyushu and Formosa in the first ever long-range bombing missions flown over the ocean, to bomb the Chinese cities of Nanking and Nan-ch'ang. These missions delivered a shock to "…Western complacency about Japanese progress in air power." [38] Over the following months the G3M bombers of the First and Second Combined Air Groups flew repeated missions against Chinese air bases and cities down the Yangtze River; carrier bombers participated as well. These missions were unescorted, as the Japanese Navy's available fighters lacked the range, and the land-based bomber units and carrier bombers suffered accordingly. The arrival of the Type 96 Fighter (A5M) to bases around Shanghai with the transfer of land-based air units from Japan helped address the balance, enabling the Japanese Navy forces to establish a measure of air superiority and escort the bombers on all but their longest range missions.[39]

Despite its large territorial conquests, culminating in the successful capture of the city of Hankow in October 1938, the Japanese Army was unable to force the Chinese Government into surrendering. The Japanese high command decided to launch a sustained aerial attack on China's cities and lines of communication in an attempt to demoralize the Chinese population and strangle supplies to the Chinese armies.[40] Between 1938 and 1941 the Japanese Navy's G3M bombers pounded the Chinese capital at Chungking and dozens of other cities, flying repeated missions from their bases around Hankow, sometimes with the support of the Army's new Type 97 bombers (Ki-21). Up until the late summer of 1940 these missions were unescorted, the A5M lacking the range to accompany the bombers all the way to Chungking. Even with larger formations allowing greater defensive fire, the G3M units suffered losses at the hands of intercepting Japanese fighters up until the advent of the Navy's newest fighter, the Type 0 Carrier Fighter, the superlative Mitsubishi A6M Reisen.[41] Carrier air units with the B5N and the D3A bombers were used to attack Chinese lines of communication up and down the China coast. While the Navy did manage to maintain air superiority over its area of operations in central and southern China, the strategic air campaign against Chinese cities failed to alter the military situation, which remained stalemated.[42]

On the eve of the Greater East Asia War Japan's naval aviation units, both land and carrier based, were far and away the most capable among the world's leading navies.[43] Japan's naval pilots had gained a wealth of experience from the air war in China, were equipped with some of the finest naval airplanes in the world, and had perfected their doctrine and combat tactics. With the expiration of the Washington Naval Treaty in 1936 Japan was free to build more aircraft carriers without limitations, adding the *Shokaku* and her sister ship the *Zuikaku* a few months before the outbreak of war in the Pacific, giving the Navy a total of 10 aircraft carriers. In addition to converting more carrier and land-based fighter units to the

A6M, the finest naval fighter of its time, the Navy introduced what would become the second iconic Japanese naval aircraft of the Pacific War, the Type 1 Land-based Attack Bomber, and the Mitsubishi G4M, shortly to become infamous among Allied soldiers, sailors, and airman as the "Betty". At the beginning of December 1941 the Japanese Navy had some 1,800 carrier and land-based airplanes poised for attack across the Pacific. But embedded within this powerful force were certain fatal flaws that would not become apparent until months later.[44] First and foremost, the Japanese Navy's strategy for the Pacific War, its doctrine, and its equipment were all geared to and predicated on a short decisive conflict. The Navy's airplanes achieved their performance, notably their exceptional range, at the cost of lighter structural weight and the absence of protection for their crews or fuel tanks, as well as load carrying capacity.[45] In a short, victorious war this would have been less of an issue, but in the massive battle of attrition that the Pacific War turned into the inherent weaknesses in the Japanese Navy's airplanes, like their Army counterparts, would prove to be catastrophic, given the inability of Japan's aviation industry, despite prodigious efforts, to manufacture airplanes in the quantities needed. The propaganda issued as Japan marched inexorably toward war, including the aviation postcards of the time, gave the impression of Japan's overwhelming strength in the air; it would prove to be an illusion.

The Development of Civil Aviation

During the second half of the 1930s, Japanese pilots undertook a number of long-distance flights on behalf of civilian organizations. These led to Japan's first international aviation records. In the first of these flights, in the spring of 1937 the *Asahi Shimbun* sponsored a goodwill flight between Tokyo and London for the coronation of HM King George VI. The newspaper purchased the second prototype of the Type 97 Command Reconnaissance Airplane (Ki-15) from the Mitsubishi Aircraft Company. Stripped of its military equipment and renamed *Kamikaze* (Divine Wind), the airplane left Tokyo on April 6, 1937, with Masaaki Iinuma as pilot and Kenji Tsukagoshi as mechanic and navigator, arriving in London on April 9th having covered 9,542 miles with an actual flying time of 51 hours, 17 minutes, and 23 seconds.[46] The British aviation magazine *Flight* commented that "Japan has surprised everybody, and for that reason deserves all hearty congratulations. All the world pictured her as striving with her own typical ingenuity and pertinacity to develop an aircraft industry, an aero engine industry, and a class of competent pilots. Those efforts aroused admiration, but did not attract any particular attention. No one expected that suddenly and almost without warning two Japanese airmen in a machine designed and built in Japan, driven by a Japanese engine, would make a long distance flight across the two continents of Asia and Europe..."[47] Two years later, between May 13 and 15, 1938, a long-range monoplane that had been under development by the Aeronautical Research Institute for the previous six years set a world distance record. Built by the Tokyo Gas & Electrical Industry Company Ltd. (Tokyo Gasu Denki Kogyo K.K.), the Gasuden Koken Long-range Research Aircraft flew 7,239 miles over a triangular course in 62 hours, 22 minutes, and 49 seconds.[48] The Army provided a pilot, Major Yuzo Fujita, from the Army's Air Technical Research Institute. In the summer of 1939 the *Mainichi Shimbun* announced that it was sponsoring a goodwill flight around the world. The newspaper purchased a civil transport conversion of the Mitsubishi Type 96 bomber (the G3M2 model) which Mitsubishi had developed for Japan Air Lines (Nippon Koku K.K.). Named *Nippon*, the aircraft, registered J-BACI, took off from Tokyo's Haneda Airport on August 26, 1939, flying to the United States via Alaska, and then down through Central and South America, across the Atlantic Ocean to Africa, and then on to Rome. The outbreak of war in Europe cancelled planned visits to London, Paris, and Berlin. The airplane returned to Tokyo via India and Southeast Asia, arriving at Haneda on October 20th after a flight of 32,845 miles.[49]

The 1930s saw the steady expansion of commercial aviation in Japan, both domestically and internationally. During the 1929-30 year, Japan Air Transport carried 2,755 passengers over its 2,575 km route system; by the 1938-39 year these figures had grown to 69,268 passengers and a 15,335 km route system.[50] Much of the expansion was related to Japan's military conquests on the mainland of Asia. On September 26, 1932, the Manchuokuo Air Transport Company (Manshu Kokuyuso K.K.) was organized to provide air services in Manchuokuo with Fokker Super Universal and F.7b/3M aircraft borrowed from Japan Air Transport, linking up with Japan Air Transport's services at Darien and Mukden.[51] On October 8, 1935, Japan Air Transport inaugurated a weekly passenger service between Japan and Formosa with Fokker F.7b/3M airplanes, adding a route from Osaka to Shanghai the following year.[52] Japan Air Transport's routes followed the Japanese Army's military successes in China. Between April 1938 and March 1939 the airline opened new routes between Tokyo and Peking, and between Fukuoka on the island of Kyushu and Nanking in China.[53] In view of the growing importance of air transportation to Japan's expansion abroad, the Japanese Government established a national organization to unify and control air routes between Japan, China, and Manchuokuo through the merger of Japan Air Transport and the newer International Air Transportation Company. The new company, the Greater Japan Air Lines Company (Dai Nippon Koku K.K.), came about through the Japan Air Lines Company Law promulgated in April 1939, with a capital of one hundred million yen, in which the Government retained a 37% ownership.[54] By 1940 the Greater Japan

Air Lines Company was offering a one-day schedule between Tokyo, Manchukuo, Peking, Nanking, and Canton, with plans to extend passenger services to Bangkok, Thailand.[55] Domestic expansion saw the inauguration of a night mail service between Tokyo and Osaka in November 1933 which evolved into regular passenger services, the development of express passenger services between Tokyo and Fukuoka to connect with flights to Formosa and Korea, and the progressive extension of services north to Sapporo, on the island of Hokkaido. To replace the elderly Fokker aircraft, Japan Air Transport acquired twin-engine Airspeed Envoys built under license by Mitsubishi, and Douglas DC-2s built by Nakajima; the Mitsui Busan Trading Company imported 20 Douglas DC-3s between 1937 and 1939, and during 1938 Japan Air Transport acquired 30 Lockheed Model 14 aircraft which were used on the routes between Japan and China.[56] Japan Air Transport had also asked the Mitsubishi Company to design a commercial transport for its international routes. This airplane went into service with the Greater Japan Air Lvines Company as the MC-20-I and II (the Army ordered a substantial number as the Type 100 Transport (Ki-57)).[57]

The Japanese aviation industry went through its own transformation. Toward the end of the 1930s, with the nation embroiled in the war in China and with the international situation worsening, the Government, under military influence, sought to place the Japanese economy on a war footing. The National Mobilization Law of March 1938 gave the Government "...almost unlimited powers to issue ordinances for the control of practically every part of the national economy."[58] That same month the Government introduced the Aircraft Manufacturing Enterprise Law, putting all aircraft manufacturers with a capital greater than three million yen under government control and authorizing the fourteen companies covered in the law to manufacture airplanes, engines, and accessories.[59] The purpose of the law was to give the Government greater control over the aviation industry and to stimulate production by expanding the production facilities of existing manufacturers and allowing selected new entrants into the industry.[60] Several companies, such as Aichi Aircraft, Kawasaki Aircraft, Ishikawajima Aircraft, and Hitachi Aircraft, were spun off from their parent manufacturing companies so that they could focus on airplane production, while the major manufacturers Nakajima, Mitsubishi, Kawasaki, and Tachikawa, which would collectively produce two-thirds of Japan's military aircraft during the Pacific War, received direct government support to expand their manufacturing facilities.[61] Between 1937 and 1941 Japanese aircraft production increased 3.3 times.[62] As impressive as this achievement was, like the aircraft the industry was building for the Japanese military, it masked certain fundamental weaknesses. While it is difficult to comprehend today after decades of Japan's postwar success in manufacturing, in the 1930s, unlike their American counterparts, Japan's aviation manufacturers, and much of its industry, were still coming to grips with the process of mass production; manufacturing efficiency was a little over a third of that of the United States aviation industry.[63] The industry lacked depth in the number of aeronautical engineers and skilled workers employed and the quality of aeronautical research supporting the industry, particularly with regard to airplane engines, where Japanese developments lagged the West. Worse for the industry's efficiency was the bitter rivalry between the Japanese Army and Navy air arms. The unwillingness of the Army and Navy to agree on standardized engines, armament, accessories, or even aircraft led to an unnecessary duplication of effort in production and undue strain on engineering resources.[64]

Postcards from the Years of Independence
Army Aviation

飛行先發の
勇士

我が陸軍精銳機青島李村に着陸

Japanese Army Air Force pilots with their Kawasaki Army Type Otsu 1 Reconnaissance aircraft at the beginning of the Manchurian Incident September-October 1931.

Kawasaki Army Type 88-1 Reconnaissance aircraft on an airfield in China during the Manchurian Incident, with Nakajima Army Type Ko 4 fighters in the background. Both types were used during the conflict in Manchuria and the fighting around Shanghai in February 1932.

（日支交戰滿洲事變）　平壤より奉天に到着の日本飛行機

The Nakajima Army Type 91 Fighter replaced the elderly Army Type Ko 4 Fighter toward the end of the fighting in Manchuria. Nakajima built 350 Type 91 Fighters and Ishikawajima Aircraft built an additional 100 or more.

Although its overall performance was somewhat disappointing, in 1932 the Army began to introduce the Mitsubishi Army Type 92 Reconnaissance Aircraft, which saw service at the tail end of the Manchurian Incident and in the early phase of the Sino-Japanese War.

On January 10, 1932, the Army received its first patriotic donations of so-called "Aikoku" aircraft, a Junkers K-37 and a Dornier Do-B Merkur configured as an ambulance airplane. The practice of raising funds to purchase airplanes for the Army continued well into World War II.

（日本）ユンカース愛國第1號機 674

Aikoku 1, the Junkers K-37, at Yoyogi Parade Ground in Tokyo during the presentation ceremony. Shortly after the airplane went to Manchuria.

愛國五十（熊本）號の雄姿

陸軍省

A representative package of postcards presented to donors during the presentation ceremony for an Aikoku aircraft, in this case a gift from the citizens of Kumamoto Prefecture, on the island of Kyushu.

A typical presentation ceremony. The Minister of War christens the new airplane "Tomikuni", named for the company that donated the Army Type 91 Fighter to the Army, then presents the documents to the president of the company, who in the bottom photo stands next to the fighter with a member of his Board of Directors.

荒木陸軍大臣の「富國」命名

根津社長獻納目錄捧呈

「富國」號と根津社長・吉田重役

A postcard showing the front of a Type 93 Heavy Bomber (Ki-1). The corrugated panels show the influence of Junkers designs. The Type 93's slow speed and heavy handling did not endear it to its crews.

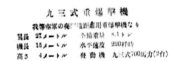

The Mitsubishi Army Type 93 Heavy Bomber (Ki-1), one of the transitional aircraft the Army acquired during the mid-1930s that continued in service until replaced with the later Army Type 97 Heavy Bomber (Ki-21).

The crew of a Type 93 Heavy Bomber (Ki-1) prepare for a mission. Note the metal steps built into the wing so the crew can climb to the cockpit.

聯備り出發ヲ待ツ

（飛行第七聯隊）　　機　行　飛

娘子關を越えて
重爆隊

The Mitsubishi Army Type 93 Twin-Engine Light Bomber (Ki-2), which entered service at the same time as the Army Type 93 Heavy Bomber. The airplane proved popular with its crews for its maneuverability, and like its heavier brother saw service in the early stages of the Sino-Japanese War.

The Army Welfare Department issued this postcard showing a squadron of the later model Army Type 93-2 Heavy Bombers (Ki-1-2) flying over north China. A rough translation of the first line of the caption is "Going over the checkpoint". The artist was O. Toda, and the card was published by the Deko Printing Company.

The Kawasaki Army Type 93 Single-Engine Light Bomber (Ki-3), which began to replace the Army Type 88 Light Bomber in 1934. The postcard describes the Type 93 as efficient, fast, and nimble while bombing, but engine problems plagued the type during its short period of operational service.

九 二 式 戰 闘 機
B.M.W. 600馬力（水冷式）

The Kawasaki Army Type 92 Fighter. Although the Army accepted 385 Type 92 Fighters to serve alongside the Type 91 Fighter, it was not as popular with Army pilots.

A Nakajima Army Type 94 Reconnaissance Aircraft (Ki-4) about to pick up a message from an Army unit during the fighting around Shanghai in the fall of 1937.

Battle line of Shanghai 1937 Japanese plane carrying correspondence tube to the front

The Japanese Army went to war in China in the summer of 1937 with the airplanes the Army had accepted in the mid-1930s. Here two Type 92 Reconnaissance Aircraft fly over the Chinese town of Nanyuan, east of Wuhan.

検閲濟

嵐を衝いて南苑爆撃の我が飛行機

Japanese planes attack Nanyuan.

支那事變 持輯ニユース

わが荒鷲部隊

支那事變 特輯ニュース
無敵空軍

壯絕爆彈投下の我が無敵陸軍機

済閲檢　　　　Bombing of invincible Japanese army air force

An Army Type 93 Single-Engine Light Bomber on a bombing mission over northern China during 1937.

支那事變 特輯ニュース
わが荒鷲部隊

銀翼を張つて石家莊爆擊に向ふ我が海軍機

済閲檢　　　Japanese naval planes on their way te attack Shihchiachuang

Although labeled as a Japanese Navy airplane, this postcard shows an Army Type 93 Heavy Bomber on its way to bomb Shihchiachuang.

The Mitsubishi Army Type 97 Command Reconnaissance Aircraft (Ki-15) had exceptional performance for its time, being virtually immune to interception from almost all Chinese Air Force fighter planes. The Army Ministry (Rikugunshō) issued this postcard with the inscription that the airplanes shown "are doing well at the front lines".

「號本日全」機納獻がられわく輝勳武に空の場戰

A tinted postcard showing three Type 97 Command Reconnaissance Aircraft in loose formation flying over China.

陸軍航空本部貸下　陸軍新重爆撃機

The Army's new Mitsubishi Army Type 97 Heavy Bomber (Ki-21), which made its maiden flight in December 1936.

The Japanese Army Air Force released this postcard of the Mitsubishi Army Type 97 Heavy Bomber (Ki-21). The Type 97 Heavy Bomber entered the war in China in the fall of 1938 with the 60th Sentai, where it replaced the elderly Type 93 Heavy Bomber.

陸軍航空本部貸下

陸軍新輕爆擊機

As part of its modernization program the Army ordered the Mitsubishi Army Type 97 Light Bomber (Ki-30), which went into service in 1938. When operating within range of Japanese fighters the Type 97 Light Bomber proved to be a reliable and effective aircraft.

陸軍航空本部貸下

陸軍新輕爆擊機

A postcard taken from a photo released by the Army showing two Type 97 Light Bombers flying in formation. The description simply states that this is the Army's new light bomber; postcards rarely indicated the specific designation.

Despite having better flying qualities than its Type 97 Light Bomber counterpart, the Kawasaki Army Type 98 Single-Engine Light Bomber (Ki-32) failed to find favor with the Army due to the Type 97 Light Bomber's more reliable radial engine.

陸軍航空本部貸下

陸軍新輕爆擊機

陸軍省　(軍事郵便)　獻納 愛國機

漢口陷落記念
昭和13年10月28日

In the fall of 1938 the Army Postal Service issued a set of commemorative postcards which included this card of a Type 98 Single-Engine Light Bomber presentation "Aikoku" aircraft given to the Army on October 28, 1938.

陸軍省　戰 鬪 機　愛國第四五四 (東京産組)

The Nakajima Army Type 97 Fighter (Ki-27) boasted phenomenal maneuverability, reflecting the Army fighter pilots' obsession with individual dogfighting. With two 7.7mm machine guns, the Type 97's armament was no different from the fighter planes of World War I. This airplane was the Army's 454th presentation aircraft.

A row of Army Type 97 Fighters (Ki-27) line up for a review.

陸 軍 戰 闘 機

Both the Army Type 97 Heavy Bomber (Ki-21) and the Army Type 97 Fighter (Ki-27) were heavily involved in the fighting over the Nomonhan. The Type 97 Fighter pilots found themselves hard pressed against the faster and more heavily armed Polikarpov I-16 fighters, whose pilots learned to use hit and run attacks against the more maneuverable Japanese fighters.

The Army accepted the Kawasaki Army Type 99 Twin-Engine Light Bomber (Ki-48) in 1940. This postcard shows a formation of Type 99 Twin-Engine Light Bombers from the 2nd Chutai of the 45th Sentai, which featured a red tail marking.

Naval Aviation

In 1931 the Hiro Naval Arsenal built the Navy Type 90-1 Flying Boat (H3H1), the first large all-metal airplane designed by Japanese aeronautical engineers without foreign assistance. While not a success, the Type 90-1 provided valuable experience in building large flying boats for the Navy.

Taken from a well-known painting of the event, this postcard shows the combat between the American Robert Short in his Boeing Model 218 and the Type 13 Carrier Bombers and Type 3 Carrier Fighters of the Imperial Japanese Navy around Shanghai on February 22, 1932.

The Nakajima Navy Type 90 Carrier Fighter (A2N1). This airplane is a "*Hokoku*" aircraft, as the Navy's airplanes donated through patriotic contributions were called. This airplane is named "Ichikwawa-Go".

The Yokusaka Navy Type 92 Carrier Attack Aircraft (B3Y1), designed to replace the Navy Type 89 Carrier Attack Aircraft. This airplane is also a presentation aircraft, donated to the Navy with funds raised through a savings bank.

The Nakajima Navy Type 90-2-2 Reconnaissance Seaplane (E4N2). Another "*Hokoku*" aircraft, named "Tomikuni-Go", after the company that donated the airplane to the Navy. This is the same company that donated the Type 91 Fighter to the Army shown in the postcard above.

The Yokosuka Navy Type 96 Carrier Attack Bomber (B4Y) saw action during the Sino-Japanese War and remained in first line service until 1940. This postcard was one of a series the Navy Ministry's Department of Public Relations printed showing the Navy's aircraft.

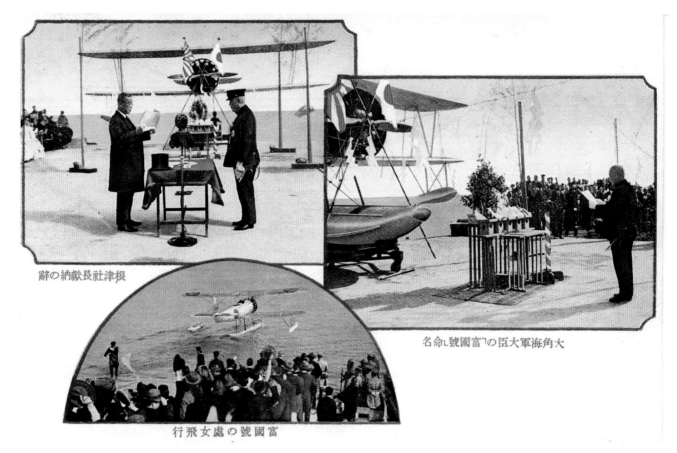

辭の納獻長社津根

大角海軍大臣の「富國號」命名

富國號の處女飛行

The presentation ceremony for Navy Type 90-2-2 "*Hokoku*" No. 20. On the top left, the president of the company presents the airplane to the Navy. On the right, the Navy Minister formally christens the airplane, which taxis away in the photo on the lower left.

海軍省軍事普及部御貸下

艦上攻撃機

This postcard of a Navy Type 90 Carrier Fighter was published by the Lion Toothpaste Company as one of a Greater Japan Flying Boys Collection for children. The fighter was donated to the Navy with money raised through children's donations around the country.

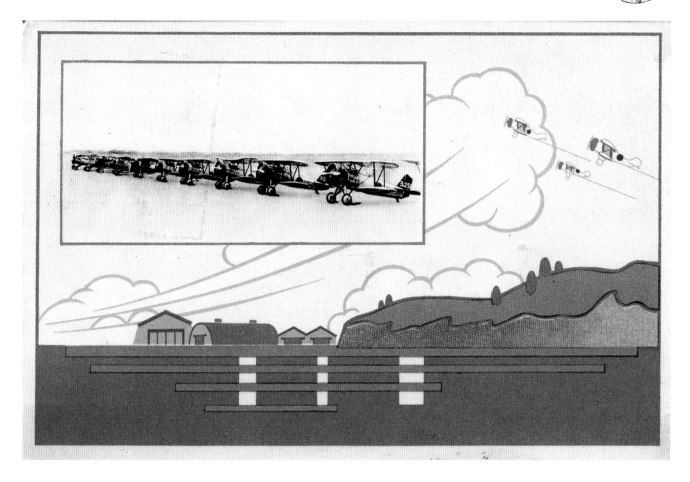

A line of Navy Type 95 Carrier Fighters against a representation of a Navy airfield, with three fighters zooming into the sky on the right.

The Nakajima Navy Type 95 Carrier Fighter (A4N1). Entering service in 1936, the Type 95 replaced the older Type 90 Carrier Fighter, though it too was soon replaced in first line service. Armed with two 60kg bombs, the Type 95 fighters flew close support missions in the early months of the Sino-Japanese War.

The Aichi Navy Type 96 Carrier Bomber (D1A2). The Type 96 Carrier Bomber was an interim type, a bridge between the older types and the more modern all-metal monoplane aircraft that entered service a short time later. The Type 96 achieved a certain amount of fame as the aircraft that bombed the U.S. Navy's gunboat *Panay*. This is a presentation aircraft (No.150), donated to the Navy by the Fukuoka Prefecture Industry Union.

A postcard from the same Navy Ministry series showing the Kawanishi Navy Type 94 Reconnaissance Seaplane (E7K2), which remained in service until 1942 on convoy escort and anti-submarine patrol duties.

報國第二〇二號　（第一高雄號）〔陸上攻擊機〕
報國第二〇三號　（第二高雄號）　同　型

The Mitsubishi Navy Type 96 Land-Based Attack Bomber (G3M1). Entering service in 1936/37, the Type 96 demonstrated exceptional range, and in the opening weeks of the Sino-Japanese War flew the first trans-ocean bombing missions in history. The two aircraft shown here are both "*Hokoku*" presentation aircraft (No.202 and No.203) named "Ko-o-Go", which loosely translates as "Hope of Winning". The censor appears to have removed any trace of the Type 96 upper gun turrets.

The Mitsubishi Navy Type 96 Carrier Fighter (A5M1). With the Navy Type 96 Land-Based Attack Bomber and Type 96 Carrier Fighter, the Japanese aviation industry could be said to have achieved parity with the West. The Type 96 Carrier Fighter was the world's first monoplane carrier fighter.

The A2M2 version of the Navy Type 96 Carrier Fighter featured a more powerful engine and a three-bladed propeller. This aircraft was one of two Type 96 Carrier Fighters (Nos. 494 and 498) donated to the Navy by the Japanese Patriotic Wives Association.

Three Navy Type
96 Carrier Fighters
fly above the flag
of the Imperial
Japanese Navy.

A postcard taken from a painting of the first trans-oceanic bombing raid flown by the Type 96 Land-Based Attack Bombers of the Kanoya Kokutai from Formosa on August 14, 1937, to bomb targets in China. The Kisarazu Kokutai flew a similar mission the next day from Kyushu in their Type 96 bombers. The artist has labeled the painting "penetrating the dark and heavy clouds" for the weather encountered en route.

A postcard showing the Type 96 Land-Based Attack Bombers of the Kisarazu Kokutai fighting their way through heavy weather en route to bomb Nanking on August 15, 1937. The postcard was taken from a painting by an artist named Tanaka and was published by the Maitsuru Navy Hall Pavilion.

In August 1937 the Kisarazu Kokutai had on strength eight Hiro Naval Arsenal Navy Type 95 Land-Based Attack Aircraft (G2H1). When losses of the Type 96 Land-Based Attack Aircraft over Nanking mounted, the Type 95s were pressed into service, operating from the island of Saishuto (Cheju Do), off Korea. This is a postcard of a painting depicting the eight Type 95s heading off on a bombing mission to China, though in actuality one airplane crashed on the first mission on September 30, 1937.

The Aichi Type 99 Carrier Bomber (D3A1). In this postcard the Type 99's landing gear has been edited for some reason. The Type 99 went into service in 1940 and was the Imperial Navy's premier dive bomber in the first years of the Pacific War. This presentation airplane (No.524) is named "Zen Nippon-Go" (All Japan), and appears to be the 57th airplane donated to the Navy through this group or organization.

報國 第五二四號 （第五七 全日本號） 艦上機

支那事變 特輯ニュース

わが荒鷲部隊

南京爆撃出動準備なり 晴れの首途

津門檢   Receiving instructions before leaving to bomb Nanking.

During the first few months of the Sino-Japanese War the Japanese Government published a series of postcards with Japanese and English captions showing Japanese soldiers, sailors, and airmen in action. This postcard shows Imperial Navy Type 96 Land-Based Attack Bomber crews being briefed for a mission to the Chinese capital at Nanking in the fall of 1937.

支那事變 特輯ニュース
わが荒鷲部隊

勇躍爆撃に出勤せんとする我が荒鷲部隊

済閲検

Japanese air force ready for an attack.

Both the carrier *Akagi* and the *Kaga* operated the Navy Type 96 Carrier Bomber during the early years of the Sino-Japanese War. In the fall of 1937 the *Kaga's* air group flew the Type 96 Carrier Bomber (D1A2) on close support missions for the Japanese Army, earning the Army's praise for the accuracy and effectiveness of their attacks.

支那事變 特輯ニュース
わが荒鷲部隊

快翔する荒鷲我が海軍機の勇姿

済閲検

Gallant Japanese naval planes.

A postcard from the same series showing a camouflaged Navy Type 96 Land-Based Attack Bomber heading off on a mission. Note how the censor has removed any indication of the upper gun turrets.

新鋭式六九海軍機

The two Navy Type 96 Carrier Fighters (A5M) shown here are pictured in the markings of the Dai 12 Kogun Kokutai, a land-based air group that flew a mix of fighters, dive bombers, and attack bombers from bases in China.

A postcard commemorating a famous incident from the Sino-Japanese War. On December 9, 1937, Warrant Officer Kanichi Kashimura of the 13th Kokutai, flying a Type 96 Carrier Fighter (A5M1), collided with a Chinese Air Force Curtiss Hawk III during a dogfight over Nanchang. Kashimura lost a third of his port wing but managed to fly safely back to his base. This postcard, showing Kashimura's Type 96 fighter on its return to base, was issued to commemorate the exhibition of his airplane in Tokyo.

A Navy Type 95 Reconnaissance Seaplane (E8N2) being lowered on to the sea for a mission. The Type 95 served aboard the Imperial Navy's seaplane tenders, battleships, and cruisers. During the Sino-Japanese War the Type 95 was used for reconnaissance and artillery spotting.

壮途に就かんとする水上機

（海軍省梗下）

Although not in pristine condition, this dramatic postcard of a Kawanishi Navy Type 94 Reconnaissance Seaplane (E7K1) surrounded by the rising sun emblem of the Japanese Navy celebrates the capture of the Chinese city of Wuhan in October 1938 as part of what the inscription calls, in a rough translation, the "long-term peace building in the East".

支那海封鎖に躍動する海軍新鋭艦上攻撃機

昭和十七年八月十一日
海軍省許可濟第四七二號

Several of the Navy's land-based air groups employed the Nakajima Navy Type 97 Carrier Attack Bomber (B5N1) in China as level bombers from late 1938 through 1941 flying in support of the Japanese Army.

我が海軍の精鋭

渡洋空襲（中攻）

（不許複製）検閲濟

This postcard of a formation of Navy Type 96 Land-Based Attack Bombers was issued in October 1940 in commemoration of the Navy's first trans-oceanic raid in August 1937. In this photo the censor has left visible one of the Type 96's upper turrets.

The superlative Mitsubishi Navy Type 0 Carrier Fighter (A6M2), which transformed the air war over China following its combat debut in September 1940 and dominated the skies across the Pacific in the early months of the Greater East Asia War. This is a "*Hokoku*" presentation aircraft (Number 646) named "Orimono-go", possibly for a textile company. The relatively low presentation number could indicate that the presentation aircraft was purchased during 1941. Unfortunately there is no date to indicate whether the Navy was willing to release a photo of its latest fighter at that time.

Civil Aviation

The Mitsubishi Karigane I *Kamikaze*, the civil version of the Type 97 Command Reconnaissance Airplane, which made the record flight from Tokyo to London between April 6 and 9, 1937.

The Gasuden Koken Long-range Research Aircraft that broke the World's distance record in May 1938.

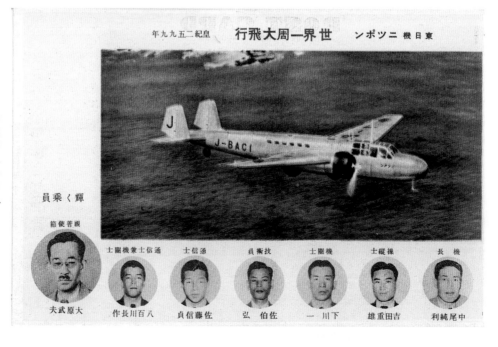

The crew of the Mitsubishi G3M2 J-BACI that flew around the world between August and October 1939. On the left is a picture of Takeo Ohara, the single passenger, who was the official Japanese envoy for the flight.

AIR-STATION OF MANCHOU AVIATION CO, HSIN-CHING,
滿洲航空會社新京區飛行場　（新　京）

A Nakajima-built Fokker
Super Universal of the
Manchuokuo Air Transport
Company (Manshu
Kokuyuso K.K.) at one of
its airfields in China.

A tinted postcard showing the colorful
markings of one of Manchuokuo
Air Transport Company's Fokker
Super Universals. After production
was transferred from the Nakajima
Aeroplane Company, the Manchuokuo
Air Navigation Company built 35 Super
Universals in Manchuria.

Another tinted postcard of a
Manko MT-1 Hayabusa Passenger
Transport, designed and built by
the Manchurian Air Transport
Company as a replacement for its
Fokker Super Universals. Production
continued at the Manchurian
Aeroplane Manufacturing Company
and later at the Nihon Kokusai
Koku Kogyo K.K., with some 50
to 60 MT-1s being built in total.

Nakajima assembled five Douglas DC-2s for Japan Air Transport to replace its Fokker F.VIIb/3M aircraft. The first aircraft was completed in early 1936 and went into service on the route from Fukuoka to Taiwan. Intended for license production in Japan, Japan Air Transport opted for the larger and more capable Douglas DC-3.

The Douglas DC-2 inspired Nakajima engineers to design a smaller passenger plane for short-range routes. The Nakajima AT-2, shown in this postcard, carried a crew of three and eight passengers. The airplane saw service with Japan Air Transport and the Manchuokuo Air Transportation Company, and as a military transport with the Japanese Army and Navy as the Army Type 97 Transport (Ki-34) and Navy Type AT-2 Transport (L1N1).

快翔する MC20 型輸送機

A dramatic shot of a Mitsubishi MC-20 Passenger Transport with Mount Fuji in the background. Based on its Army Type 97 Heavy Bomber (Ki-21), Mitsubishi's transport could carry eleven passengers and went into service in 1940 with the Greater Japan Air Line Company (Dai Nippon Koku K.K.). The Army adopted the airplane as the Army Type 100 Transport (Ki-57), and a small number went to the Japanese Navy as the Type 0 Transport (L4M1)

This unique
postcard consists
of a Japanese
woodblock print
pasted onto
a cardboard
backing. Using
the techniques
from the Japanese
Ukiyo-e tradition,
the print shows
a stylized
airplane speeding
over an ocean
freighter with
a bold pattern
of waves below.
This postcard
was issued on
October 8, 1935,
by the NYK
line (Nippon
Yusen K.K.),
Japan's largest
shipping line, to
commemorate
the beginning
of commercial
flights between
Japan and Taiwan/
Formosa.

From the same series issued by the NYK shipping line to commemorate the start of commercial flights between Japan and Taiwan/Formosa, this woodblock print postcard shows Japan and its two colonies, Korea and Formosa, with a design of a Fokker F.VIIb/3M airplane that flew along the route to Taiwan.

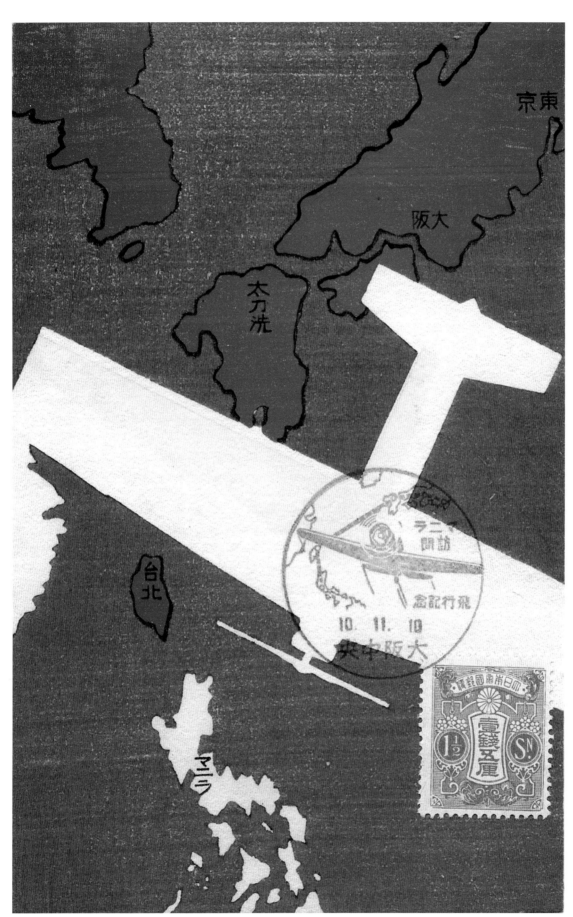

Another striking woodblock print design with the image of an airplane superimposed over a map showing the main stops on the route from Tokyo to Taipei onto Taiwan. This card was issued on November 10, 1935.

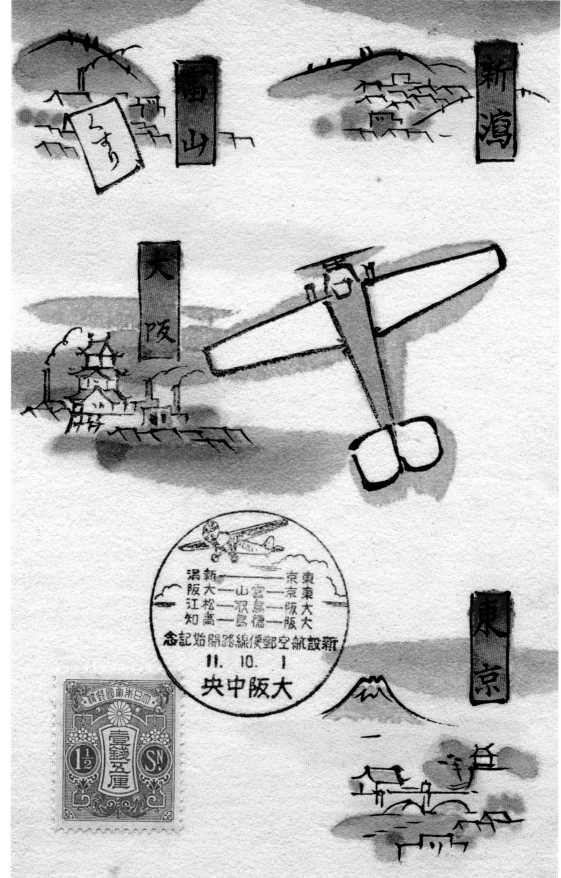

This woodblock print postcard commemorates the establishment of air mail services between Tokyo, Osaka, and several points between and beyond. Tokyo is shown in the lower right corner, represented by the view of Mount Fuji, while Osaka isv on the middle left, represented by a view of its famous castle. The postcard was issued on October 1, 1936.

Although postmarked for a patriotic flying week in June 1945, this postcard and the next are believed to be from an earlier NYK shipping line series, as it shows a Douglas DC-2 from the mid-1930s. The DC-2 flies over a pink bank of cloud with a representation of Mount Fuji rising up through the clouds.

A stylized airplane with striking colors, particularly the large rising sun emblem. Mount Fuji rises in the background.

POST CARD

ハワイ眞珠港爆撃の海鷲　　　　　　　　　　高井貞二畫伯筆

CHAPTER FOUR

The Greater East Asia War

1941 – 1945

"Now the Enemy is America and Britain!"

O N DECEMBER 8, 1941, RADIOS ACROSS JAPAN announced that war with Britain and the United States had begun. The Imperial Japanese Army and Navy had launched attacks across Asia, striking at Pearl Harbor, Hong Kong, and the Philippines. Newspapers carried the Imperial Rescript Declaring War. vBritain and America, the Emperor told his subjects, were "menacing gravely the existence of Our Empire."[1] To one middle school boy the news was electrifying: "I felt as if my blood boiled and my flesh quivered. The whole nation bubbled over, excited and inspired. 'We really did it! Incredible! Wonderful!' That's the way it felt then."[2] For years the Japanese people had

been fed a steady diet of propaganda about the superiority of Japan and Japan's sacred mission to liberate Asia from the yoke of Western imperialism. Now, one after another, the bastions of colonial rule fell to the Japanese Army and Navy: Hong Kong, Manila, Singapore, Batavia (as Jakarta was then called), and Rangoon. In mere six months Japan transformed the face of Asia. In those heady early days few gave much thought to the reckoning that was to come.

The Japanese Army Air Force and Naval Air Force had been at the forefront of Japan's march across Southeast Asia. The speed of the advance and the capability of Japanese air power came as a shock to the Allied powers. In the years leading up to the outbreak of war in the Pacific, the Japanese military had been

"...shallowly evaluated; they were underrated at best, despised at worst."[3] Japanese military aviation was seen as markedly inferior to the Western powers, the Japanese deemed incapable of matching the West. In an article on Japanese air power for the September 1941 issue of the American magazine *Aviation*, the associate editor disparaged Japanese airplanes as mere copies of their Western counterparts. "Japanese imitativeness", he wrote, "...notoriously extends to copying other countries' blueprints of aircraft and aero engines, this circumstance accounting for the obsolescence of their flying machines."[4] Moreover, he wrote that "... Japanese fliers are utterly inexperienced in the mass tactics of air war as it is being fought in Western and Eastern Europe."[5] While correctly identifying the inherent weakness of Japan's industrial base, Western observers assumed that Japanese military capabilities were equally far behind. The British magazine *The Aeroplane* wrote in March 1941 that "Japanese aviation has, therefore, a long way to go before it will be able to compete successfully with, or even combat, the 'decadent European and American democracies'".[6]

In part, Western ignorance was a direct result of the Japanese military's ability to keep its latest aircraft a secret, even from the Japanese people. For example, reports of the success of the Type 0 Carrier Fighter (A6M1) in the skies over China during 1940-41 simply referred to Japan's "Sea Hawks" without mentioning the type of aircraft involved.[7] The existence of the Type 0 Carrier Fighter was not made public until well after the outbreak of the Greater East Asia War. The aviation postcards published just before the war and over the two years following the outbreak of the war seldom, if ever, referred to a specific model aircraft. The Japanese Army Air Force would sometimes identify an airplane as a fighter, or bomber, or reconnaissance aircraft, while the Japanese Navy often resorted to simple euphemisms, such as "wild eagles" or "brave Navy eagles". But once war had begun postcards showing the Army and Navy aircraft involved

in achieving Japan's victories became popular. Both the Army and the Navy produced several series of postcards showing the principal aircraft employed in the early part of the war. Japan's early victories became a source of inspiration for the hundreds of artists the Army and Navy employed to produce patriotic paintings and sculptures to inspire the public. Many of these works were displayed at the First Greater East Asia War Art Exhibition in Tokyo in December 1942 and the Second Greater East Asia War Art Exhibition later in 1943.[8] The Army and the Navy produced postcards of many of these paintings, showing Japanese aircraft victorious in battle with Allied ships and airplanes. Other war art series gave the Japanese public a view of life at the front lines, showing the day to day work of the airmen and the ground crews. The paintings were done in a mix of styles, from traditional Japanese-style painting (nihonga) to Western-style (yōga). Invariably the artist attempted to portray the superior strength and spirit of the Japanese airman. As Japan's attempt to mold the Greater East Asia Co-prosperity Sphere got underway with its inherent commitment of Japan's protection, Japanese propaganda sought to portray the strength of Japan's military. The Army and Navy printed postcards in Southeast Asian languages for distribution to the local population.

The campaigns of the Pacific War will be well known to most readers and need no elaboration in this volume. The Battle of Midway in June 1942 was the first check on Japanese military expansion. Thereafter, the Japanese Army and Navy underwent a grinding battle of attrition in the Southwest Pacific. As American industrial production got into full swing, the flow of weapons, troops, and supplies enabled the Allies to push Japan's defensive barrier back toward the home islands through offensives in the Central and South Pacific, crippling the Japanese Navy in the process. While Japan's aviation industry managed prodigious feats of production during the war, the rate of attrition was so high and

America's production advantage so great that the Japanese Army Air Force and Naval Air Force were simply overwhelmed. The Japanese Navy had attacked Pearl Harbor with six aircraft carriers and some 400 airplanes; in the battle for Okinawa in 1945 the U.S. Navy marshaled a task force with fifteen aircraft carriers and over 1,000 airplanes. The development of more capable aircraft for the Army and Navy, with better armament, more powerful engines, and better protection, came too late in the war to have an influence. Japan's airplane production peaked in 1944 and declined rapidly thereafter.[9]

From mid-1943, as the tide of war turned against Japan, the tone of official Japanese propaganda began to change. The Japanese word for "fight to the death" began to replace the word for "decisive battle". There were fewer and fewer real victories to celebrate, though the Japanese press continued to report what were in fact imaginary victories that moved ever closer to Japan. More critically, as the American air and submarine offensive against Japan's supply lines became more effective, shortages of all forms of raw materials grew steadily. During 1944, as paper became scarce, the Government arranged the merger of many newspapers and periodicals; the 13,556 active periodicals registered in 1941 declined to 942 in 1944.[10] By that time, too, color printing disappeared for all but a small number of children's publications.[11] From circumstantial evidence, the production of airplane postcards appears to have shrunk dramatically after 1944.[12] No doubt for security reasons, the fighters and bombers of the later war years do not seem to have featured in airplane postcards. Not surprisingly, in terms of airplane postcards these last, desperate years of struggle are poorly documented.

Postcards from the Greater East Asia War

Army Aviation

爆撃の目的地は迫れり。自信満々のわが重爆撃機群　　　　　　　　　　　　陸軍航空本部検下

This postcard of a Type 97 Heavy Bomber (Ki-21) is from a series the Japanese Army issued in July 1942. The caption reads "Getting closer to the target a group of confident heavy bombers".

A postcard from the same July 1942 series showing the Type 1 Fighter (Ki-43 Hayabusa) which entered combat in December 1941. The caption describes the Type 1 as the Army's new fighter that is "… showing great effort in every direction in the southern sky".

南方の空に縦横の活躍を展開する新戦闘機隼號　　　　　　　　　陸軍航空本部検下

制空戰線

Postcard sets were often issued in a special envelope, sometimes with striking graphics. This envelope was for a set of postcards of Japanese Army Air Force aircraft published in July 1942. The postcards in this set were in black and white; the same set of postcards was also issued in color, as illustrated in the next postcard.

南方〇〇基地に於けるわが陸軍新戦闘機隼號の待機

陸軍航空本部貸下

The Army issued a similar series of postcards in color in July 1942, using some of the same images, under the title "Victory of the Army Eagles". This postcard shows two Type 1 Fighters, possibly from the 64th Sentai, in Malaya or Thailand earlier in 1942. The caption reads "The Army's new fighter planes standing by on the field at the southern ------ base".

A postcard of a Type 1 Fighter Hayabusa. The caption says that this is the Army's new fighter plane scouting for enemies on the way to its destination.

索敵しつゝ目的地に向ふ新戦闘機隼號

陸軍航空本部貸下

堂々南方の空を制壓する陸軍新鋭輕爆撃機

陸軍航空本部貸下

This postcard shows a formation of Type 99 Light Bombers (Ki-48) on a mission, possibly over China.

重
疊
た
る
天
嶮
の
上
空
を
壓
し
つ
ゝ
飛
翔
す
る
輕
爆
擊
機

陸軍航空本部貸下

Two Type 99 Light
Bombers from the 45th
Sentai flying over China.

敵
上
空
悠
々
作
業
す
る
新
銳
偵
察
機

陸軍航空本部貸下

A postcard from the Army's July 1942 series showing a Type 99 Assault Bomber (Ki-51). This airplane appears to be from the 3rd Chutai of the 27th Sentai, which fought in the Malayan campaign and in China. For some reason the Army's censors did not remove the Sentai markings. The caption states that this is the new reconnaissance aircraft that is working in enemy skies.

颯爽基地を離陸せんとする重爆撃機

The Army censor did not remove the Sentai markings from this Type 97 Heavy Bomber. The stripes on the tail indicate the aircraft is from the 3rd Chutai of the 12th Sentai, which operated over Burma in the first months of 1942.

南方の基地に悠々機種を誇る戦闘機

A row of Type 97 Fighters (Ki-27s). The caption reads "The proud fighter planes calmly waiting at the base in the south." The Army sensor appears to have removed the Sentai markings from the first three aircraft in line, but the fourth aircraft, with green camouflage, carries the markings of the 50th Sentai. The 50th Sentai's Type 97 Fighters fought in the Philippines and then over Burma. The 50th Sentai began re-equipping with the Type 1 Fighter in February 1942, which would indicate that this scene was likely in Thailand, where the 50th Sentai was based to support the invasion of Burma.

重爆の整備に忙殺される整備員

Army maintenance crews working on the engine of a Type 97 Heavy Bomber.

The Army was proud of its latest fighter and issued several postcards showing the Type 1 Fighter Hayabusa.

世界に冠たる我が陸軍最新鋭戦闘機『隼 號』の勇姿

陸軍省検閲済

わが陸鷲の至寶

新鋭戦闘機 „隼"

マレーに、ジャバに „隼" の羽ばたくところ赫々たる戦果に輝く。すばらしい速度、たくましい戦闘性能。„隼" 戦闘部隊の名はいまや世界の驚異である。　陸軍航空本部許可済

This postcard was printed by the Aviation Boys Club. The title reads: "Our Army Eagle's Treasure: the New Hayabusa Fighter Plane". The card explains that in Malaya and Java, wherever the Hayabusa flies it shines in battle. "Amazing speed, strong fighting power. The Hayabusa unit's reputation is now 'the World's Wonder'".

省 軍 陸　　　　　（商魚京東）〇九八第國愛　機 闘 戰

During the war the Army continued the practice of accepting patriotic donations of "Aikoku" aircraft. This Type 1 Fighter Hayabusa, shown in the markings of the 50th Sentai, is Aikoku aircraft number 890. It was donated to the Army by a group of fishing companies from the Tokyo area.

陸軍新鋭戰闘機「隼」號

Another postcard of the same aircraft. The Hayabusa was a beautifully proportioned airplane. The markings of the 50th Sentai, which returned to Japan to re-equip with the Hayabusa, can just be seen on the rear fuselage.

省軍陸 〔院議衆〕一三三二第國愛 機闘戰

A postcard from sometime later in the war showing Aikoku aircraft number 2331, a Type 1 Fighter Hayabusa donated by the Lower House of the Diet, the Japanese parliament.

省軍陸 （釦スレプ崎石）四三二二第國愛 機闘戰

This shows Aikoku aircraft number 2234, a Type 2 Fighter (Ki-44) Shoki donated to the Army by the Ishizaki Press.

Type 99 Light Bombers (Ki-48) in flight. These aircraft appear to be from the 2nd Chutai of the 3rd Sentai, based in the Kurile Islands in 1942. The caption says "light bombers impressively flying through the sky".

堂々空を壓する輕爆機   陸軍省許可濟

峻峰を瞰下して飛翔する重爆編隊   陸軍省許可濟

A formation of Type 97 Heavy Bombers flying over China. The postcard notes that this photo was approved for release by the Army.

Type 97 Heavy Bomber crews receiving instructions before a flight. Note that the caption, which refers to "Japanese eagles", has been translated into Indonesian.

Garoeda Nippon melawan pesawat pelèmpar bom moesoeh.

テキ ヲ バクゲキ ニ ムカフ ニツポン ノ アラワシ

ニッポン バクゲキキ ノ ヘンタイヒカウ

Rombongan pesawat pelèmpar bom Nippon terbang dengan rapi.

A card probably from the same series with the caption in Indonesian showing a formation of Type 97 Heavy Bombers in flight.

One of the few postcards that refers to an Army airplane by its official name, in this case the Donryu (Storm Dragon), the name for the Type 100 Heavy Bomber (Ki-49). This postcard was issued with the approval of the Army.

呑 龍   陸軍省航空本部許可濟

ただ一弾

小松崎茂筆

During the war, the Japanese Army employed hundreds of artists and illustrators for propaganda purposes. This postcard, published by the War Ministry, shows a Type 1 Fighter Hayabusa shooting down a Royal Air Force Hurricane fighter. The title of the painting on which it was based is "Just one shot!", and the artist was Komatsuzaki Mamoru.

This is a postcard of a painting by the artist Konosuke Tamura titled "Air Raids on Rangoon, Burma", published by the Army Museum Association. The painting depicts an air battle over Rangoon, with Japanese Army Air Force fighters engaged in combat with Royal Air Force Hurricanes, Buffalos, and a lone Spitfire (which did not serve in Burma until later in 1943) diving away on the lower left. The artist painted a Type 1 Fighter Hayabusa in the middle of the picture to honor the spirit of the late Lt. Col. Takeo Kato, the famous commander of the 64th Sentai killed in action over Burma.

ラングーン爆撃

（陸軍省許可濟）　　　ビルマ方面陸軍派遣　　田村孝之介

A painting of a gunner firing on attacking fighters. The slogan on the postcard reads "Find the enemy and destroy them!" This postcard was part of a series showing the Japanese Army and Navy in action with British and American forces.

A postcard from the same series depicting an air battle between Japanese fighter planes and American fighters and bombers.

豪快！　急降下爆撃

陸軍派遣画家　吉岡堅二筆

Another postcard published by the Army Museum Association showing a painting by the artist Kenji Yoshioka. Army bombers are diving down on their target. The caption reads "Splendid! Glorious! Making a steep dive to attack below."

This postcard is part of an extensive series of paintings by a number of artists showing the activities of Japanese Army airmen and their ground crews. This postcard, of a painting by the artist Shin Kurihara, shows pilots getting a lesson in tactics.

This postcard from the same series shows two ground crew changing the tire on a Type 97 Light Bomber (Ki-30).

A radio operator on an Army bomber transmits a message back to base.

An artist named Teizō Itoh did this painting of a Type 97 Heavy Bomber crew receiving last minute instructions.

伊藤悌三畫

A painting of two student pilots receiving instruction.

藤田嗣治畫

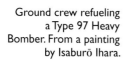

Ground crew refueling a Type 97 Heavy Bomber. From a painting by Isaburō Ihara.

伊原宇三郎畫

宮本三郎

待機

第二十八回二科美術展覧會出品

From a painting titled "Waiting", an experience that the pilots and crews of all the combatants would have recognized. This was from a painting by Saburo Miyamoto which was entered in one of the Greater East Asia War Art exhibitions.

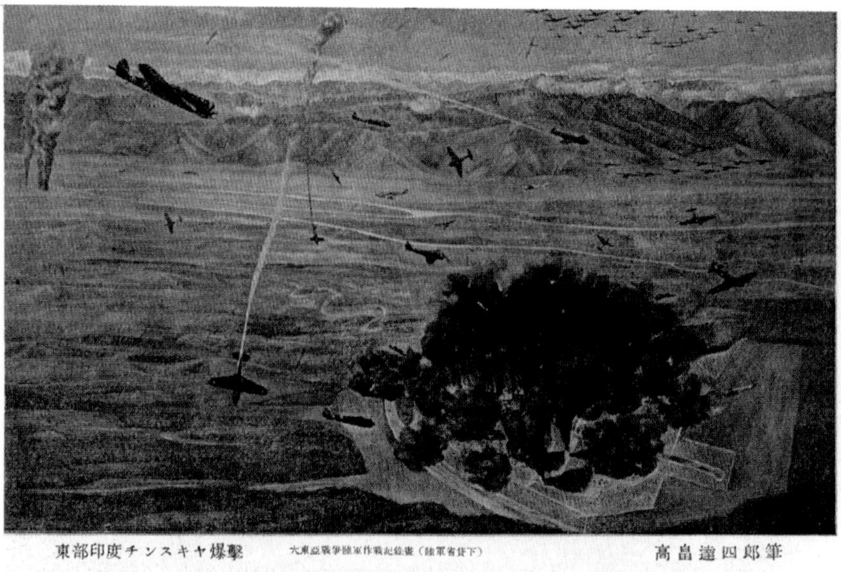
東部印度チンスキヤ爆撃
大東亞戰爭陸軍作戰記錄畫（陸軍省貸下）
高畠達四郎筆

A postcard from later in the war, showing a painting by the artist Tatsuhirō Takabatke depicting an air battle between Type I Fighter Hayabusas and American P-40s over Assam, in India.

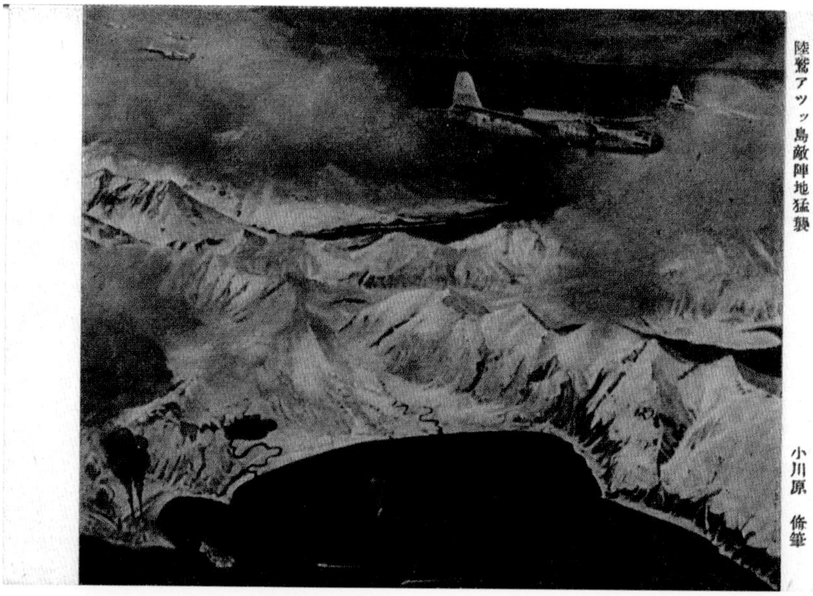
陸鷲アッツ島敵陣地猛襲
小川原 脩筆

This postcard of a painting by Shun Ogawara titled "Fierce attack on the enemy's base" shows a formation of Army Type 100 Heavy Bombers flying over Attu Island, in the Aleutians, which in fact these aircraft never did.

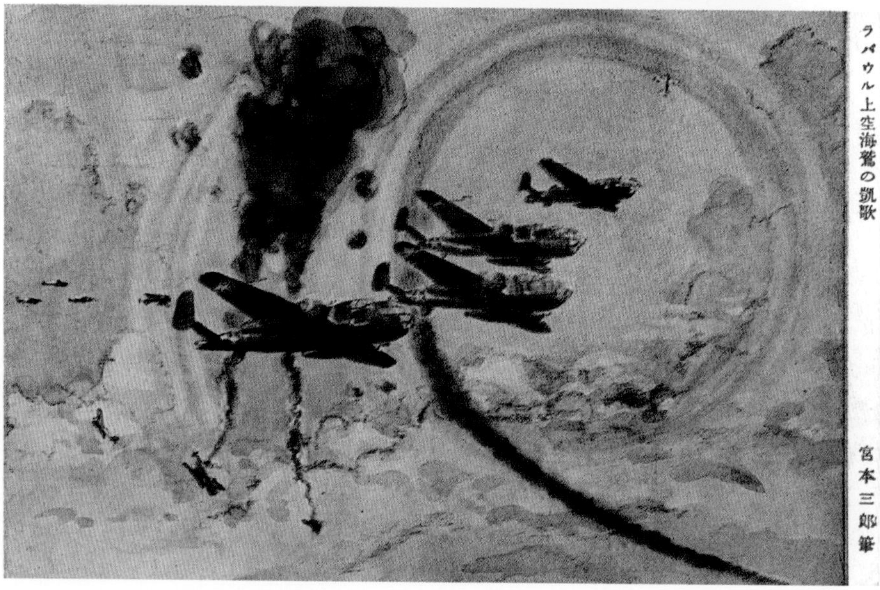
ラバウル上空海鷲の凱歌
宮本三郎筆

This postcard of a painting by the artist Saburo Miyamoto was published by the Army Art Association. It shows Japanese fighters shooting down American B-25 bombers over Rabaul. The title translates roughly as "Victory song of the Rabaul defenders".

暁を衝いて

陸軍省許可濟

Two Type 97 Heavy Bombers crossing a coast somewhere in the war zone. A rough translation of the caption reads "Bombers are piercing the dawn".

夜間着陸

陸軍省許可濟

A Type 97 Fighter (Ki-27) taxies in after a mission. The white band under the port wing indicates that this airplane was from a fighter Sentai on Defense of Japan duty. Several fighter Sentais based on the home islands used the Type 97 Fighter well into 1943.

This postcard is titled simply "Air battle", and shows the gunner on a Type 97 Heavy Bomber preparing to fire on the fighters diving from above.

空中戰鬪   陸軍省許可濟

Naval Aviation

During 1942 the Imperial Navy approved the release of a number of postcards showing Navy aircraft in action with patriotic slogans. This postcard shows a formation of Type 96 Land Based Attack Bombers on a mission. The caption reads "The planes that continue to dominate the Southern Skies with their wings."

A Type 0 Carrier Fighter (A6M2) taking off from the deck of a Japanese carrier, possibly the *Shokaku* or *Zuikaku*, in 1942. The caption reads "Wild eagles flying off the mother carrier to attack."

The Japanese Navy did not immediately release photographs of the Pearl Harbor attack. The first photos of the attack came out on January 7, 1942. A week later, on January 13, 1942, a set of postcards showing photos of the attack came out in this striking envelope. The top postcard shows a Type 97 Carrier Attack Bomber launching from one of the aircraft carriers in Admiral Nagumo's striking force. The lower postcard shows a Type 97 Carrier Attack Bomber flying over Pearl. The caption reads "Under our silver wings the Pearl Harbor base area after the attack."

横隊案を漂つて目的地に向ふわが海軍新鋭攻撃機   海軍省許可済第二二二號

A close-up shot of a Type I Attack Bomber flying in formation.

This postcard, from the same series as the one above, shows two Type 97 Carrier Attack Bombers practice bombing.

一彈必殺爆撃に戦果を上るわが海の荒鷲   海軍省許可済第二二二號

わが無敵海軍空襲部隊の新鋭攻撃機   海軍省許可済第二二二號

This postcard, from the same series as the previous card, shows a formation of Navy Type I Attack Bombers (G4M1). The caption reads "Our invincible Navy bomber unit's latest attack plane."

The Japanese would often send seasonal greetings to their friends and relatives. This postcard, apparently number eight in a series, is a patriotic greeting card for the summer season. The card shows the crew of a Type 96 Land Based Attack Bomber. The extensive caption on the right describes the crew working hard on their way to the target. "Finally we reach the enemy's land", the card reads. "The big bombing attack is about to commence."

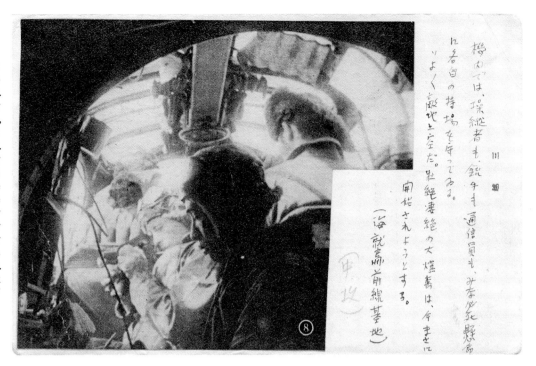

画郎次松上村　　　　　戦海イワハ

The Navy, like the Army, commissioned artists and illustrators to produce patriotic paintings and illustrations celebrating the Navy's victories. This painting, by the artist Matsugiro Murakami, titled "Hawaii Sea Battle" shows the artist's impression of the attack on battleship row at Pearl Harbor.

This postcard of a painting by the artist Takeru Matsuzoe, published by the Navy Welfare Group, depicts the attack on the *Prince of Wales* on December 10, 1941, by the Type 96 Land Based Attack Bombers of the Genzan and Mihoro air groups of the Twenty-second Air Flotilla. The artist has taken some liberty in his depiction of the attack, as the two air groups actually bombed the battleship from an altitude of 5,000 feet.

クラークフィールド攻撃   佐藤　敬書

海軍省貸下   朝日新聞社主催大東亞戰爭美術展覧會

キヤビテ軍港攻撃   三輪晁勢書

海軍省貸下   朝日新聞社主催大東亞戰爭美術展覧會

Titled "Raid on Clark Field", this postcard of the painting by the artist Satō Kei shows the devastating attack on the morning of December 8, 1941. The painting featured in the book *Greater East Asia War Navy Art* and was displayed at an exhibition organized by the Greater East Asia War Exhibition Association, hosted by the *Asahi Shimbun* newspaper.

Published by the *Asahi Shimbun* for the Pacific War Exhibition Association, this postcard shows a painting by the artist Chōsei Miwa depicting the Japanese Navy's attack on the U.S. Navy's base at Cavite, near Manila, on December 8, 1941. The painting was titled "Attack on the Naval Port of Cavite, Manila".

This postcard depicts the sinking of the *Prince of Wales* and the *Repulse* with torpedoes carried by the Type 1 Attack Bombers of the Kanoya Air Group on December 10, 1941. The original painting was titled "The Battle of the Malayan Sea" by the artist Kenichi Nakamura, and appeared in the First Greater East Asia War Art Exhibition in the fall of 1942.

大東亞戰爭美術展覽會出品

（マレー沖海戰）

海軍省貸下

中村研一作

This postcard shows a painting of the Imperial Navy's carrier attack on British Naval ships in the Indian Ocean in April 1942. The Navy's Type 99 Carrier Dive Bombers (D3A1) sank one British aircraft carrier, two heavy cruisers, and two destroyers. This painting was entered in the Second Greater East Asia War Art Exhibition in 1943.

The Japanese Navy saw the Battle of the Coral Sea as a clear victory. This postcard shows the torpedo attack on the USS *Lexington*, which was lost during the battle.

Another artist's interpretation of the Japanese Navy's attack on the USS *Lexington* during the Battle of the Coral Sea, showing Type 97 Carrier Attack Bombers dropping torpedoes in the *Lexington*'s path.

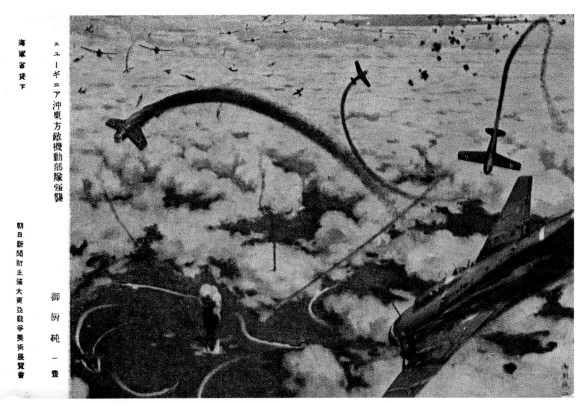

ニューギニア沖東方敵機動部隊強襲

海軍省貸下

御厨純一画

朝日新聞社主催大東亜戦争美術展覧會

"Attacking the Enemy Fleet Off New Guinea", the artist Junichi Mikurga's impression of an air battle between Japanese Navy aircraft and U.S. Navy F4F Wildcats seen going down in flames. Unusually, this painting shows a Japanese bomber diving down in flames. The painting was exhibited at the Greater East Asia War Art Exhibition.

小堀安雄 海軍報道班員 イサベル島沖海戦 第二回大東亜戦争美術展覧會出品 海軍省貸下

The painting represented in this postcard was titled "Battle of Cape Isabel". It depicts Navy Type 99 Carrier Dive Bombers in the Battle of the Eastern Solomons, off Santa Isabel Island, in August 1942. This painting was done by Yasumasa Kobori, an official Navy artist, and was an entry in the Second Greater East Asia War Art Exhibition in 1943.

新井勝利 山路口報海軍班 整備作業に於ける 航空母艦上に於けるに整備作業 第二回大東亞戰爭美術展覽會出品 海軍省貸下

Navy war artists also painted scenes of everyday activities in addition to battle scenes. This postcard, which has elements of the more traditional "Nihonga" style of painting, is from a painting showing Type 0 Carrier Fighters warming up on the deck of an aircraft carrier. The title of the painting was "Maintenance on a carrier" by the artist Katsumari Arai. This painting was also one of the entries in the Second Greater East Asia War Art Exhibition.

山口華楊 日本報海軍班 基地に於ける整備作業 第二回大東亞戰爭美術展覽會出品 海軍省貸下

A Navy reporter/artist did this painting of a Navy Type 2 Floatplane Fighter (A6M2-N) unit at an island base somewhere in the South Pacific. On the completion of a mission, a ground crewman carries a pilot back to the shore to be welcomed by his comrades, while maintenance crews swarm over his airplane to prepare it for its next combat. This painting was also an entry into the Second Greater East Asia War Art Exhibition.

第二回大東亞戰爭美術展覽會出品

加藤榮三 日本報海軍班 設營隊の活躍 海軍省貸下

A maintenance group working to repair an airfield, with the wreckage of an American airplane prominently in the foreground. One might interpret the painting as showing a victorious Japan consolidating its new-found position in Asia against the backdrop of an utterly defeated America. This painting featured in the Second Greater East Asia War Art Exhibition.

南方海軍航空基地

鶴田吾郎筆

海軍省許可済

Members of a Navy bomber squadron wave to their squadron mates as they take off for a mission. This was a tradition in Japanese Navy squadrons. The title of the painting on which this postcard is based was "Southern Navy airfield", and the artist was Goro Tsuruda. The postcard was published by the Navy.

南海青空の下 航空母艦より 飛上らんとする海鷲

昭和十八年三月十五日海軍省許可濟第二七〇號

During 1943 there were few victories to celebrate. The Navy continued to approve the release of photographs and postcards showing the Navy's aircraft in operation. This photo of a Type 99 Carrier Dive Bomber, released on March 15, 1943, carried the caption "Under the blue skies of the southern ocean the Navy eagle is about to fly off the carrier."

昭和十八年三月十五日海軍省許可濟第二七〇號

蒼茫の海上を大擧編隊組んで爆撃に向ふ海鷲群

During March 1943, the Navy approved the publication of a set of colored postcards showing various Navy aircraft. This postcard shows a formation of Type 96 Land Based Attack Bombers, ostensibly about to bomb their target, though by this date the Type 96 had mostly been withdrawn from combat.

飛沫ヲ揚ゲテ（適性飛行）　土浦海軍航空隊

横鎮第二七號ノ一〇四ノ二　昭和十七年十月九日許可濟

A postcard from the same March 1943 series showing Navy Type 93 Intermediate Trainers taxiing out for a practice flight. Remarkably, the postcard identifies these airplanes as belonging to the Tsuchiura Kaigun Kokutai (air group), one of the Japanese Navy's principal bases for the Naval Preparatory Flight Training Program (Yokaren), located 34 miles northeast of Tokyo in Ibaraki Prefecture.

海鷲 の 勇姿　　　　　　海軍省第八三五號許可濟

An interesting postcard from the March 1943 series showing the later Type 99 Carrier Dive Bomber Model 22, which entered service in the fall of 1942. The caption says "The brave look of the Navy eagle".

"Wild eagles going forth" is the title of this postcard showing a formation of Type 2 Floatplane Fighters, probably published during 1943.

荒
鷲
前
進

海軍省第八三五號許可濟

艦隊の眼となる新鋭水上偵察機は曉天を衝いて索敵に向ふ

昭和十八年三月十五日
海軍省許可濟第二七〇號

Although not pristine, this postcard is included because of its comparative rarity. The card shows a Navy Type 1 Reconnaissance Seaplane (E13A2), one of the less common Navy aircraft to appear in postcards.

整備兵の聲に送られて出撃する新鋭陸上攻撃機

昭和十八年三月十五日海軍省許可濟第二七〇號

Ground crew waving off a Type 1 Attack Bomber as it sets off on a mission.

Endnotes

Introduction

1. Westney, D. Eleanor: *Imitation and Innovation: The Transfer of Western Organizational Patterns to Meiji Japan* (Cambridge: Harvard University Press, 1987), P.134.

2. Morse, Anne Nishimura: "Art of the Japanese Postcard", in *Art of the Japanese Postcard: The Leonard A. Lauder Collection at the Museum of Fine Arts, Boston* (Boston: Museum of Fine Arts, 2004), P.15.

3. Morse, "Art of the Japanese Postcard", P.15.

4. Morse, "Art of the Japanese Postcard", pp.17-18.

5. Brown, Kendall H.: "Postcards, Commerce, and Creativity in Japan, 1904-1940", in *Art of the Japanese Postcard: The Leonard A. Lauder*
Collection at the Museum of Fine Arts, Boston (Boston: Museum of Fine Arts, 2004), P.48.

6. Hallion, Richard P.: *Taking Flight: Inventing the Aerial Age from Antiquity through the First World War* (New York: Oxford University Press, 2003), P.264.

7. United States War Department, Military Intelligence Service: *Soldier's Guide to the Japanese Army* (Washington, D.C.: Government Printing Office, 1944), P.33.

8. Brown, "Postcards, Commerce, and Creativity in Japan, 1904-1940", P.49.

9. Jansen, Marius B.: *The Making of Modern Japan* (Cambridge, MA: Harvard University Press, 2000), pp.600-605; Beasley, W.G.: *The Modern History of Japan, Third Revised*

Edition (Tokyo: Charles E. Tuttle, 1973), pp.236-42.

10. Earhart, David C.: *Certain Victory: Images of World War II in the Japanese Media* (Armonk, NY: M.E. Sharpe, Inc., 2008) P.xi.

11. Earhart, *Certain Victory*, P.82.

12. Kushner, Barak: *The Thought War: Japanese Imperial Propaganda* (Honolulu: University of Hawaii Press, 2006), P.6.

13. Kusher, *The Thought War*, P.45.

14. Brown, "Postcards, Commerce, and Creativity in Japan, 1904-1940", P.62.

Chapter One
The Pioneering Years, 1910-1918

1. Peattie, Mark R.: *Sunburst: The Rise of Japanese Naval Air Power, 1909-1941* (Annapolis, MD: Naval Institute Press, 2001), P.3.

2. Peattie, *Sunburst*, P.4; Kohri, Katsu, et al: *Aireview's The Fifty Years of Japanese Aviation 1910-1960, Book Two* (Tokyo: Kantosha Co. Ltd., 1961), pp.4-5.

3. *Aireview's The Fifty Years of Japanese Aviation*, P.5.

4. *Aireview's The Fifty Years of Japanese Aviation*, P.6.

5. *The Japan Times*, March 13, 1911.

6. *The Japan Times*, May 11, 12, June 2, 1912.

7. Peattie, *Sunburst*, P.4.

8. *Aireview's The Fifty Years of Japanese Aviation*, P.8; *The Japan Times*, November 12, 13, 1912.

9. Mikesh, Robert C. and Shorzoe Abe: *Japanese Aircraft 1910-1941* (London: Putnam Aeronautical Books, 1990), P.46.

10. Mikesh and Abe, *Japanese Aircraft 1910-1941*, P.53.

11. *Aireview's The Fifty Years of Japanese Aviation*, P.13; *Flight*, October 9, 1914, P.1026.

12. *Aireview's The Fifty Years of Japanese Aviation*, pp.13-14; Sekigawa, Eiichiro: *Pictorial History of Japanese Military Aviation* (London: Ian Allen Ltd., 1974), P.11; Mikesh and Abe, *Japanese Aircraft 1910-1941*, P.48.

13. Mikesh and Abe, *Japanese Aircraft 1910-1941*, pp.52-3, 55.

14. Mikesh and Abe, *Japanese Aircraft 1910-1941*, P.55.

15. Davilla, Dr. James J. and Arthur M. Soltan: *French Aircraft of the First World War* (Stratford, CT: Flying Machine Press, 1997), pp.397, 471; Fujiwara, Hiroshi and Toshio Fujita: *Baron Miyahara and His World of Aircraft: Military and Civil Aircraft 1910-1970* (Tokyo: Japanese Aeronautic Association, 2008), P.67.

16. Davilla, Dr. James J. and Arthur M. Soltan: *French Aircraft of the First World War*, pp.491, 498.

17. Mikesh and Abe, *Japanese Aircraft 1910-1941*, P.52.

18. Peattie, *Sunburst*, pp.6-7; Mikesh and Abe, *Japanese Aircraft 1910-1941*, P.263; *Aireview's The Fifty Years of Japanese Aviation*, P.13.

19. Peattie, *Sunburst*, pp.8-9; *Aireview's The Fifty Years of Japanese Aviation*, pp.13-14; *Flight*, November 13, 1914, P.1118.

20. Peattie, *Sunburst*, P.14.

21. Mikesh and Abe, *Japanese Aircraft 1910-1941*, pp.264-65.

22. Mikesh and Abe, *Japanese Aircraft 1910-1941*, P.269.

23. Mikesh and Abe, *Japanese Aircraft 1910-1941*, P.268.

24. Peattie, *Sunburst*, pp.9-10.

25. Davilla, Dr. James J. and Arthur M. Soltan: *French Aircraft of the First World War*, P.531.

26. *Aireview's The Fifty Years of Japanese Aviation*, pp.9-12.

27. *Aireview's The Fifty Years of Japanese Aviation*, pp.10-13, 19-20.

28. *Aireview's The Fifty Years of Japanese Aviation*, P.18.

29. *Aerial Age Weekly*, January 29, 1917, P.503.

30. Mikesh and Abe, *Japanese Aircraft 1910-1941*, pp.198-200; *Aireview's The Fifty Years of Japanese Aviation*, P.21.

Chapter Two
The Apprenticeship Years, 1919-1930

1. Beasley, *The Modern History of Japan*, P.214.

2. Beasley, *The Modern History of Japan*, pp.215-17; Jensen, *The Making of Modern Japan*, pp.528-34.

3. League of Nations: Conference for the Reduction and Limitation of Armaments: *Armaments Year-book 1924* (Geneva: League of Nations, 1924), P.571; *Armaments Year-Book 1930*, P.599.

4. Boyd, Carl: "Japanese Military Effectiveness: The Interwar Period", in Millett, Allan R. and Williamson Murray: *Military Effectiveness: Volume 2: The Interwar Period*, New Edition (Cambridge: Cambridge University Press, 2010), P.135.

5. Humphreys, Leonard A.: *The Way of the Heavenly Sword: The Japanese Army in the 1920's* (Stanford, CA: Stanford University Press, 1995), P.92.

6. Samuels, Richard J.: *"Rich Nation, Strong Army": National Security and the Technological Transformation of Japan* (Ithica, NY: Cornell University Press, 1994), pp.109-10.

7. Samuels, *"Rich Nation, Strong Army"*, P.110.

8. Samuels, *"Rich Nation, Strong Army"*, pp.112-13.

9. *Aireview's The Fifty Years of Japanese Aviation*, P.22; Sekigawa, *Pictorial History of Japanese Military Aviation*, P.16; Katushi, Owaki: "The 1919 French Aviation Mission to Japan, Pt.1", *Arawasi International* (April-June 2007, Issue #5); "The 1919 French Aviation Mission to Japan, Pt.2", *Arawasi International* (July-August 2007, Issue #6).

10. Mikesh and Abe, *Japanese Aircraft 1910-1941*, pp.143-44.

11. Cea, Eduardo: *Japanese Military Aircraft No.6: Bombers of the Imperial Japanese Army 1939-1945* (Valaldolid, Spain: AF Editores, 2010), pp.4, 6.

12. League of Nations: Conference for the Reduction and Limitation of Armaments: *Armaments Year-book 1927* (Geneva: League of Nations, 1927), P.580.

13. Mikesh and Abe, *Japanese Aircraft 1910-1941*, P.144; Cea, *Japanese Military Aircraft No.6: Bombers of the Imperial Japanese Army 1939-1945*, P.9.

14. Mikesh and Abe, *Japanese Aircraft 1910-1941*, pp.102, 144-48, 177-80, 213-15; *Aireview's The Fifty Years of Japanese Aviation*, P.34.

15. Peattie, *Sunburst*, pp.9-10.

16. Peattie, *Sunburst*, pp.18-19; Ferris, John: "A British 'Unofficial' Aviation Mission and Japanese Naval Developments, 1919-1929", *Journal of Strategic Studies*, Vol. 5, No.3, September 1982, pp.416-23.

17. Ferris, "A British 'Unofficial' Aviation Mission and Japanese Naval Developments, 1919-1929", P.424; Colonel the Master of Sempill, "The British Aviation Mission to the Imperial Japanese Navy", *Flight*, April 10, 1924, pp.209-13.

18. Peattie, *Sunburst*, P.19.

19. Mikesh and Abe, *Japanese Aircraft 1910-1941*, pp.161-65; Peattie, *Sunburst*, P.23.

20. Evans, David C. and Mark R. Peattie: *Kaigun: Strategy, Tactics, and Technology in the Imperial Japanese Navy 1887-1941* (Annapolis, MD: Naval Institute Press, 1997) P.332.

21. James, Derek N.: *Gloster Aircraft Since 1917*, 2nd Edition (London: Putnam Aeronautical Books, 1987), pp.157-60; Mikesh and Abe, *Japanese Aircraft 1910-1941*, pp.224-25.

22. Jackson, A.J.: *Blackburn Aircraft Since 1909*, 2nd Edition Revised (London: Putnam Aeronautical Books, 1988), pp.295-301.

23. Barnes, C.H.: *Shorts Aircraft Since 1900*, 2nd Revised Edition (London: Putnam Aeronautical Books, 1989), pp.153-54; Mikesh and Abe, *Japanese Aircraft 1910-1941*, pp.93-4.

24. Mikesh and Abe, *Japanese Aircraft 1910-1941*, pp.97-8.

25. Mikesh and Abe, *Japanese Aircraft 1910-1941*, pp.271-2.

26. Mikesh and Abe, *Japanese Aircraft 1910-1941*, pp.223-4, 273-5.

27. *Aireview's The Fifty Years of Japanese Aviation*, pp.28, 30-3, 37; Davies, R.E.G.: *A History of the World's Airlines* (London: Oxford University Press, 1964), pp.88-9.

28. *Aireview's The Fifty Years of Japanese Aviation*, pp.23, 29-30, 35-6.

29. Mikesh and Abe, *Japanese Aircraft 1910-1941*, pp.61, 124-5, 142-3, 160, 198-9; Samuels, *"Rich Nation, Strong Army"*, pp.112-14; Sekigawa, *Pictorial History of Japanese Military Aviation*, pp.21, 33.

Chapter Three
The Years of Independence, 1931-1940

1. Jensen, *The Making of Modern Japan*, pp.582-3.

2. Humphreys, *The Way of the Heavenly Sword*, P.IX.

3. Tohmatsu, Haruo and H.P. Wilmott: *A Gathering Darkness: The Coming of War to the Far East and the Pacific 1921-1942* (Oxford: SR Books, 2000), P.17.

4. League of Nations: Conference for the Reduction and Limitation of Armaments: *Armaments Year-book 1932* (Geneva: League of Nations, 1932), pp.187-92; *Armaments Year-book 1937* (Geneva: League of Nations, 1937), pp.553-57, 559-66.

5. Boyd, "Japanese Military Effectiveness: The Interwar Period", pp.138-40; *Armaments Year-book 1932*, pp.187-92.

6. Boyd, "Japanese Military Effectiveness: The Interwar Period", P.141.

7. Tohmatsu, *A Gathering Darkness*, P.60; Boyd, "Japanese Military Effectiveness: The Interwar Period", pp.141-2.

8. Samuels, *Rich Nation, Strong Army*, P.115.

9. Samuels, *Rich Nation, Strong Army*, P.115.

10. Samuels, *Rich Nation, Strong Army*, P.115.

11. *Aireview's The Fifty Years of Japanese Aviation*, pp.42-3; Sekigawa, *Pictorial History of Japanese Military Aviation*, pp.36-7.

12. *Aireview's The Fifty Years of Japanese Aviation*, pp.42-43; Thompson, Paul: "Propaganda Propellers", *Arawasi International Extra 1* (December 2009).

13. Sekigawa, *Pictorial History of Japanese Military Aviation*, pp.38.

14. Mikesh and Abe, *Japanese Aircraft 1910-1941*, pp.183-6.

15. Mikesh and Abe, *Japanese Aircraft 1910-1941*, pp.152-3, 155.

16. Sekigawa, *Pictorial History of Japanese Military Aviation*, pp.40-1.

17. Sekigawa, *Pictorial History of Japanese Military Aviation*, pp.40-1.

18. Sekigawa, *Pictorial History of Japanese Military Aviation*, pp.46-9; Francillon, René J.: *Japanese Aircraft of the Pacific War* (London: Putnam Aeronautical Books, 1987), pp.90-2, 155-63, 164-7, 196-203.

19. Cea, *Japanese Military Aircraft No.6: Bombers of the Imperial Japanese Army 1939-1945*, P.17.

20. Cea, *Japanese Military Aircraft No.6: Bombers of the Imperial Japanese Army 1939-1945*, P.27; Sekigawa, *Pictorial History of Japanese Military Aviation*, pp.64-5.

21. Kotelnikov, Vladimir R.: *Air War Over Khalkhin Gol, Air Wars 2* (Bedford, UK: SAM Publications, 2010), P.63.

22. Sekigawa, *Pictorial History of Japanese Military Aviation*, P.87.

23. See Harvey, A.D.: "Army Air Force and Navy Air Force: Japanese Aviation and the Opening Phase of the War in the Far East", *War in History*, Vol. 6, No.2, (April 1999).

24. *Aireview's The Fifty Years of Japanese Aviation*, P.43; Sekigawa, *Pictorial History of Japanese Military Aviation*, P.39.

25. Peattie, *Sunburst*, pp.27-8; Samuels, *Rich Nation, Strong Army*, pp.115-6.

26. Sekigawa, *Pictorial History of Japanese Military Aviation*, P.44.

27. Sekigawa, *Pictorial History of Japanese Military Aviation*, P.49.

28. Sekigawa, *Pictorial History of Japanese Military Aviation*, pp.52-3.

29. Peattie, *Sunburst*, pp.39-40.

30. Peattie, *Sunburst*, pp.41-2; Francillon, René J.: *Japanese Aircraft of the Pacific War* pp.268-71.

31. Peattie, *Sunburst*, pp.79-80.

32. Peattie, *Sunburst*, pp.80-1; Mikesh and Abe, *Japanese Aircraft 1910-1941*, pp.100-1, 171-2.

33. Peattie, *Sunburst*, P.86.

34. Francillon, René J.: *Japanese Aircraft of the Pacific War* pp.342-9; Peattie, *Sunburst*, pp.88-9.

35. Francillon, René J.: *Japanese Aircraft of the Pacific War* pp.411-6, 491; Peattie, *Sunburst*, P.95.

36. Peattie, *Sunburst*, pp.94-5.

37. Sekigawa, *Pictorial History of Japanese Military Aviation*, P.60; Peattie, *Sunburst*, P.104.

38. Peattie, *Sunburst*, pp.104-6.

39. Peattie, *Sunburst*, pp.110-11.

40. Peattie, *Sunburst*, pp.115-6.

41. Peattie, *Sunburst*, pp.118-9.

42. Peattie, *Sunburst*, pp.122-4.

43. Peattie, *Sunburst*, P.161.

44. Peattie, *Sunburst*, P.166.

45. Harvey, A.D.: "Army Air Force and Navy Air Force: Japanese Aviation and the Opening Phase of the War in the Far East", P.182.

46. Francillon, René J.: *Japanese Aircraft of the Pacific War*, pp.150-1.

47. *Flight*, April 15, 1937, P.360.

48. *Aireview's The Fifty Years of Japanese Aviation*, P.47; Mikesh and Abe, *Japanese Aircraft 1910-1941*, pp.90-1.

49. *Aireview's The Fifty Years of Japanese Aviation*, P.49; Francillon, René J.: *Japanese Aircraft of the Pacific War*, P.354.

50. "Civil Aviation in Japan", *The Far Eastern Economic Review*, (December 1939), P.498.

51. "Air Transportation in Japan and Manchukuo", *The Far Eastern Economic Review*, (December 1934), pp.547-8.

52. "Japanese Air Transport", *The Far Eastern Economic Review* (September 1936), P.399; "Civil Aviation in Japan", P.499.

53. "Civil Aviation in Japan", P.499.

54. "Civil Aviation in Japan", P.499; Davies, *A History of the World's Airlines*, P.192.

55. "Progress of Air Travel", *The Japan Advertiser Annual Review 1939-1940*, pp.85-6.

56. Mikesh and Abe, *Japanese Aircraft 1910-1941*, pp.196, 207-8; Francillon, René J.: *Japanese Aircraft of the Pacific War*, pp.499, 507-8.

57. Francillon, René J.: *Japanese Aircraft of the Pacific War*, pp.182-5.

58. Thresher, M.B.: "Japan's Economic Front", *The Japan Advertiser Annual Review 1939-1940*, P.1.

59. Samuels, *Rich Nation, Strong Army*, P.122; "Civil Aviation in Japan", P.499.

60. Samuels, *Rich Nation, Strong Army*, P.123.

61. Samuels, *Rich Nation, Strong Army*, P.123.

62. United States Strategic Bombing Survey: Aircraft Division: *The Japanese Aircraft Industry* (May 1947), Appendix X, P.155.

63. United States Strategic Bombing Survey: Aircraft Division: *The Japanese Aircraft Industry* (May 1947), P.4.

64. Samuels, *Rich Nation, Strong Army*, P.125, 127, 129.

Chapter Four
The Greater East Asia War, 1941-1945

1. Cook, Haruko Taya and Theodore F. Cook: *Japan at War: An Oral History* (New York: The New Press, 1992), P.77.

2. Earhart, *Certain Victory*, P.220.

3. Coox, Alvin D.: "The Effectiveness of the Japanese Military Establishment in the Second World War", in Millet, Allan R. and Williamson Murray: *Military Effectiveness: Volume 3: The Second World War, New Edition* (Cambridge: Cambridge University Press, 2010), P.1.

4. Zacharoff, Lucien: "Japanese Air Power", *Aviation*, September 1941, P.146.

5. Zacharoff, Lucien: "Japanese Air Power", *Aviation*, September 1941, P.146.

6. "Japan's Air Strength", *The Aeroplane*, March 7, 1941, P.284.

7. Horikoshi, Jiro: *Eagles of Mitsubishi: The Story of the Zero Fighter* (Seattle, WA: University of Washington Press, 1981), pp.97, 102-3.

8. Earhart, *Certain Victory*, pp.228, 395.

9. United States Strategic Bombing Survey: Aircraft Division: *The Japanese Aircraft Industry* (May 1947), P.33.

10. Earhart, *Certain Victory*, pp.99, 203.

11. Earhart, *Certain Victory*, pp.203.

12. According to Mr. Yoshio Tagaya after 1944 there were few aviation postcards available in Japan due to the paper shortages. Growing up in Tokyo he collected aviation postcards before and during the War.

Edward Young

EDWARD YOUNG HAS HAD A LIFE-LONG INTEREST IN AVIATION HISTORY. HE GRADUATED from Harvard College with a degree in Political Science. After serving as a Peace Corps Volunteer in Thailand, he attended the University of Washington, receiving an MA in Political Science. He spent eight years as a commercial banker working in New York and Hong Kong, and then joined Moody's Investors Service, the bond rating agency. During his twenty-year career at Moody's he served as an analyst and managing director with assignments in New York, London, Tokyo, and Hong Kong. He retired to Seattle, Washington, in 2004 to devote his time to writing aviation history.

In the course of living in Asia for fifteen years and traveling extensively in the region, Mr. Young developed a particular interest in the history of aviation in Asia and the history of the China-Burma-India Theater in World War II. He has a family connection with the region as well. Mr. Young's father was an exchange student in China during the 1930s, an intelligence officer with the Twentieth Air Force during World War II, and an Asian specialist, while his uncle was assigned to a Chinese tank regiment in Burma during the war and served with the OSS in China. Asian aviation and World War II in the Pacific and the CBI have been the focus of much of Mr. Young's research and writing. He is the author of *Aerial Nationalism: A History of Aviation in Thailand*, *Air Commando Fighters of World War II*, and several titles for Osprey Publications, including *Campaign 136: Meiktila 1945- The battle to liberate Burma*, *Warrior 141: Merrill's Marauders*, and *Combat Aircraft 87: The B-24 Liberator in the CBI*. He is the co-author with Louis Eltscher of *Curtiss-Wright: Greatness and Decline*. Mr. Young has contributed articles to *Air Enthusiast*, *Avions*, *Flypast*, and *Warbird Digest*. He is a regular volunteer at the Museum of Flight in Seattle and lectures on aviation history in the Seattle area.

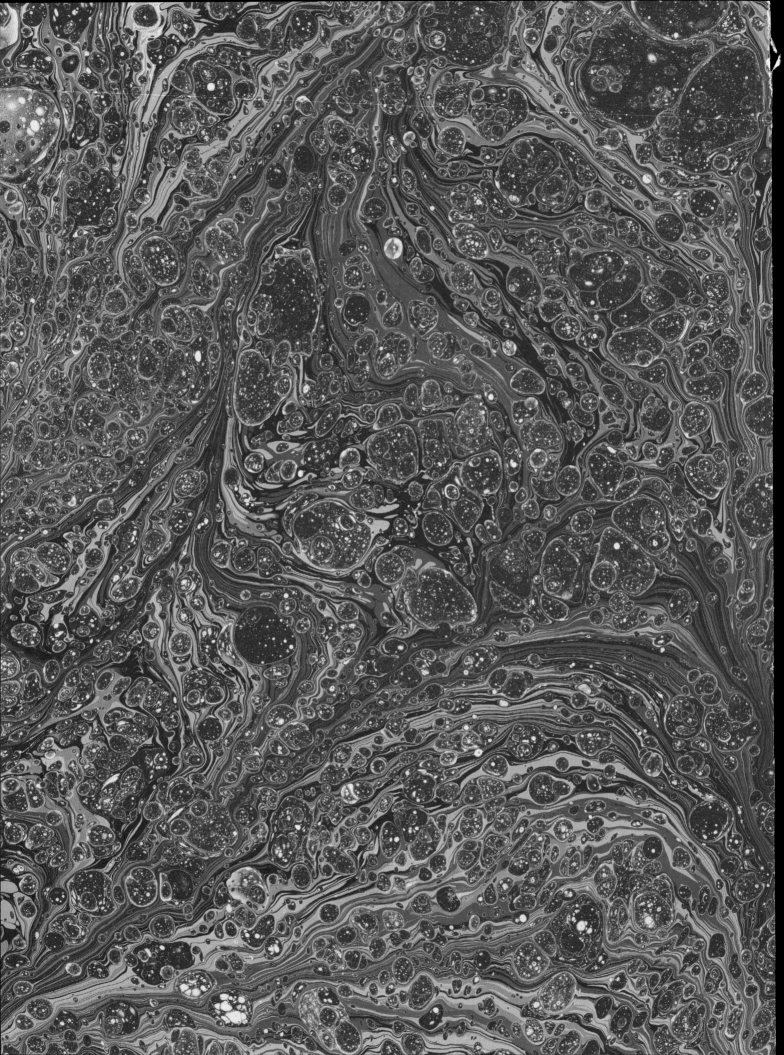